W9-BYB-586

What People Are Saying About the
Consumer Guide To Home Energy Savings

"The *Consumer Guide to Home Energy Savings* suffers from a modest title...the guide reveals how to buy and operate just about everything in your house that's connected to an electric or gas meter." — *Today's Homeowner*

"One of the best resources for homeowners that we've ever seen."
— Anne Ducey, Conservation Marketing Coordinator, Seattle City Light

"Face it, most of us don't have the inclination—or the time—to become experts... about energy efficiency and its impact on the environment. But whether we're sealing a drafty doorway, remodeling a bathroom or shopping among rows of gleaming new refrigerators, we know our actions have an impact—on the planet, as well as on our pocketbook. This book offers practical advice that will help you make smart decisions about how you build or buy.
  Not a textbook in volume or style, this paperback makes for handy toting—perhaps for perusing during your energy-efficient commute on BART or Muni."
— Beth Bourland, *The San Francisco Chronicle*

"It tells you which appliances you can replace for the biggest energy savings...in words anyone can comprehend in real sentences and nice, neat diagrams."
— *Coastal Maine News*

"A comprehensive reference guide (in plain English, no less) for homeowners who've chosen to wade through the sometimes murky waters of buying green, and money-saving, home improvement products." — *The Flint Journal*

"Not only does the *Guide* rate all the brands of major home appliances and fixtures for energy efficiency, it also shows how to insulate, landscape and maintain your home in the manner that will make your wallet and your planet greener."
— *E Magazine*

"This compact little book is full of tips on saving energy in light of costs to both you and the planet. And speaking of energy, an evening spent examining this book before you decide which major appliance to buy is a productive and efficient use of yours." — *Small Home Designs*

"If you own one book on saving energy at home, this should be it."
— Iris Communications

"An indispensable handbook for consumers who want to reduce their home energy use." — *The Consumer's Guide to Effective Environmental Choices*

"The most comprehensive guide to saving money and the environment available today." — Diane MacEachern, Tips for Planet Earth (syndicated column)

"Full of brand name evaluations to help a homeowner make more knowledgeable purchases..." — *The New York Times*

"A comprehensive, easy-to-use resource..." — *Science News*

"...could be used as a model for government information campaigns—it's an excellent, easy-to-read summary." — *Global Environmental Change Report*

"Besides up-to-date comparisons of the latest HVAC systems and appliances, the *Guide* includes an assortment of energy-saving tips." — *Fine Homebuilding*

"...comprehensive, easy-to-read prose...clear, accurate, and well-illustrated..." — NESEA's *Northeast Sun*

"Whatever your question is about home energy use, this book can almost surely answer it." — *Co-op Currents*

"...an unparalleled reference..." — *SunWorld* (International Solar Energy Society)

"With many diagrams, this is a readable, affordable guide...buy it!" — *Library Journal*

"To help maximize energy and dollar savings in your home, check out the *Consumer Guide to Home Energy Savings*....[it] is full of tips, diagrams, charts, explanations and lists on almost every facet of home life that involves energy use: insulation, windows, heating and cooling systems, food storage, cooking, dishwashing, laundry, lighting and much more. By purchasing energy-efficient appliances, you can slash your energy bills and make a positive impact on national security, the economy and the environment." — *Mother Earth News' Guide to Homes*

"Must reading for homeowners in the market for new appliances...contains a wealth of information on how to make the appliances you own now work more efficiently. The advice here will also save you hundreds of dollars a year in energy costs." — *Better Homes and Gardens*

"Don't furnish your home until you've consulted [the *Consumer Guide*]. It's a must-have resource." — *Metropolitan Home*

"There are a number of sources of information about the energy efficiency of furnaces, boilers, water heaters, and air conditioners...perhaps the most useful to the typical consumer is the [*Consumer Guide*] published by the American Council for an Energy-Efficient Economy." — *Kansas Country Living*

"You can save a lot of legwork by consulting [this book]...chock-full of tips on energy considerations of each category of appliances, as well as energy efficient tips for home renovators." — *The Green Consumer Letter*

"...a clearly-written, well-illustrated guide to insulation, air sealing techniques, heating and cooling systems, and energy-efficient lighting and appliances." — *The Journal of Light Construction*

"This book could have easily been titled *The Encyclopedia of Home Energy Savings*. It's the most comprehensive resource to home energy savings that I've seen. Every homeowner and environmentally conscious (or utility paying) renter should have a copy." — *Green Living*

SAVE MONEY, SAVE THE EARTH

CONSUMER GUIDE TO
HOME ENERGY
SAVINGS **9th** EDITION

JENNIFER THORNE AMANN
ALEX WILSON & KATIE ACKERLY

SAVE MONEY, SAVE THE EARTH

CONSUMER GUIDE TO
HOME ENERGY
SAVINGS **9th EDITION**

JENNIFER THORNE AMANN
ALEX WILSON & KATIE ACKERLY

American Council for an Energy-Efficient Economy
Washington, D.C.

NEW SOCIETY PUBLISHERS

Cataloging in Publication Data:
A catalog record for this publication is available from the National Library of Canada.

© **1990, 1993, 1995, 1996, 1998, 1999, 2003, 2007 by the American Council for an Energy-Efficient Economy. All rights reserved.**

Book design by Linda Rapp. Cover and illustrations by Katie Ackerly and David Conover.

Printed in Canada. Second printing November 2007.

Paperback ISBN: 978-0-86571-602-5

Inquiries regarding requests to reprint all or part of the *Consumer Guide to Home Energy Savings*, 9th Edition should be addressed to New Society Publishers or the American Council for an Energy-Efficient Economy at the address below.

To order directly from the publishers, please call toll-free (North America) 1-800-567-6772, or order online at www.newsociety.com

Any other inquiries can be directed by mail to the publishers at:

New Society Publishers, P.O. Box 189, Gabriola Island, BC V0R 1X0, Canada
(250) 247-9737 OR
American Council for an Energy-Efficient Economy (ACEEE)
1001 Connecticut Avenue, NW, Suite 801, Washington, DC 20036

Portions of this book have been reprinted with the permission of Massachusetts Audubon Society from the following guides: *Saving Energy and Money with Home Appliances; Oil and Gas Heating Systems; How to Weatherize Your Home or Apartment; All About Insulation;* and *Contractor's Guide to Sealing Air Leaks.*

Written comments on this publication are welcome; send your comments to:
Consumer Guide to Home Energy Savings Editor
American Council for an Energy-Efficient Economy
1001 Connecticut Avenue, N.W., Suite 801
Washington, D.C. 20036-5525
aceee.org/consumerguide

New Society Publishers' mission is to publish books that contribute in fundamental ways to building an ecologically sustainable and just society, and to do so with the least possible impact on the environment, in a manner that models this vision. We are committed to doing this not just through education, but through action. This book is one step toward ending global deforestation and climate change. It is printed on acid-free paper that is **100% post-consumer recycled** (100% old growth forest-free), processed chlorine free, and printed with vegetable-based, low-VOC inks, with covers produced using Forest Stewardship Council-certified stock. Additionally, New Society purchases carbon offsets based on an annual audit, operating with a carbon-neutral footprint. For further information, or to browse our full list of books and purchase securely, visit our website at: www.newsociety.com

NEW SOCIETY PUBLISHERS   www.newsociety.com

Contents

Lists of Tables

Acknowledgments

This book owes much to the energy conservation community: those hardworking folks who have worked diligently (and quietly) to increase the energy efficiency of the U.S. housing stock by one-third since 1973, and whose trials and errors have produced the knowledge compiled here.

The information in this guide has evolved with the steady improvements in residential appliances and changes in equipment efficiency standards. Each new edition benefits from the suggestions of experts in this field. The guide was originally co-produced with *Home Energy Magazine*, a non-profit, monthly publication covering topics ranging from research into new energy-saving technologies to tips on how to save energy in the home. Visit www.homeenergy.org/.

We thank all the individuals who have reviewed this or previous editions for clarity and technical accuracy: Carl Blumstein, Fred Davis, Neal Elliott, Howard Geller, Roger Harris, John Hayes, Drew Kleibrink, Michael L'Ecuyer, Marc Ledbetter, Karina Lutz, Chris Mathis, Alan Meier, Steve Nadel, Nancy Schalch, Mike Thompson, Hofu Wu, and Kate Offringa. Special thanks to John Morrill who contributed substantially as a co-author to past editions. We also owe special acknowledgment to the International Dark Sky Association for their contribution to the outdoor lighting section and to Harvey Sachs for authoring the new chapter on Ventilation and Air Distribution for the 9th edition. Some of the material in this chapter is adapted from the *Ventilation Guide* by Armin Rudd, Building Science Press, 2006.

Some of the illustrations and design in this book were used in or adapted from previous editions: illustrations by David Conover and layout and cover designs by Chuck Myers.

We also thank our colleagues who have contributed to the production of this guide: Harvey Sachs for his detailed review and input on updating material; Renee Nida for her copyediting and proofreading; Monica Rashkin for proofreading and indexing; and Glee Murray for her overview of the entire process. As usual, the authors remain responsible for any errors or omissions that remain.

Home Energy Checklist for Action

Here's a simple checklist to give you an idea of the things you can learn about in this book:

To Do Today

- Turn down the temperature of your water heater to the warm setting (120°F). You'll not only save energy, you'll avoid scalding your hands.

- Start using energy-saving settings on refrigerators, dishwashers, washing machines, and clothes dryers.

- Survey your incandescent lights for opportunities to replace them with compact fluorescents. These new lamps can save three-quarters of the electricity used by incandescents. The best targets are 60-100W bulbs used several hours a day. Compact fluorescent lamps (CFLs) will fit in most standard fixtures.

- Check the age and condition of your major appliances, especially the refrigerator. You may want to replace it with a more energy-efficient model before it dies.

- Clean or replace furnace, air-conditioner, and heat-pump filters.

- If you have one of those "silent guzzlers," a waterbed, make your bed today. The covers will insulate it, and save up to one-third of the energy it uses.

This Week

- Visit the hardware store. Buy a water-heater blanket, low-flow showerheads, faucet aerators, and compact fluorescents, as needed. CFLs are now sold at most drug stores and grocery stores.

- If your water heater is old enough that its insulation is fiberglass instead of foam, it clearly will benefit from a water heater blanket from the local hardware store or home supplies store.

- Rope caulk very leaky windows.

- Assess your heating and cooling systems. Determine if replacements are justified, or whether you should retrofit them to make them work more efficiently—to provide the same comfort (or better) for less energy.

This Month

- Collect your utility bills. Separate electricity and fuel bills. Target the biggest bill for energy conservation remedies.

- Crawl into your attic or crawlspace and inspect for insulation. Is there any? How much?

- Insulate hot water pipes and ducts wherever they run through unheated areas.

- Seal up the largest air leaks in your house—the ones that whistle on windy days, or feel drafty. The worst culprits are usually not windows and doors, but utility cut-throughs for pipes ("plumbing penetrations"), gaps around chimneys and recessed lights in insulated ceilings, and unfinished spaces behind cupboards and closets. Better yet, hire an energy auditor with a blower door to point out where the worst cracks are. All the little, invisible cracks and holes may add up to as much as an open window or door, without you ever knowing it!

- At night and whenever you leave your home, adjust your thermostat to save heating energy in the winter and cooling energy in the summer. Some people find it easier to install a programmable thermostat.

- Schedule an energy audit (ask your utility company or state energy office) for more expert advice on your home as a whole.

This Year

- Insulate. If your walls aren't insulated, have an insulation contractor blow cellulose into the walls. Bring your attic insulation level up to snuff.

- Replace aging, inefficient appliances. Even if the appliance has a few useful years left, replacing it with a top-efficiency model is generally a good investment.

- Upgrade leaky windows. It may be time to replace them with energy-efficient models or to boost their efficiency with weather-stripping and storm windows.

- Reduce your air conditioning costs by planting shade trees and shrubs around your house—especially on the west side.

- Know that you are making a difference!

Chapter 1

Save Money, Save the Earth

Are you about to buy a new appliance? Remodel your house? Upgrade your heating or cooling system? If you're like most of us, you don't do these things very often. When you do, you want to make good choices, both for your pocketbook and for the environment. But you probably don't have time to become an expert. That's where this book can help.

The *Consumer Guide to Home Energy Savings* will help you make wise investment decisions and help you decide which products to buy and how to use them for maximum energy savings. We've listed the best ways to tighten up your house so that your heating and cooling systems won't have to work as hard—or use as much energy. We've pulled together tips on operating new and existing appliances to reduce energy use and improve performance. But before getting into the details, let's take a look at why it makes sense to buy the most efficient appliances and conserve energy in the home.

Saving Energy—and Money—in Your Home

The wonderful thing about saving energy is that, in addition to helping the environment, you save money. It's like contributing to a good cause and ending up with more money in your pocket. Many of the energy-efficient appliances and heating or cooling systems covered in this book cost no more than their inefficient counterparts. With most others, the extra cost is easily repaid in energy savings over just a few years. To top it off, many energy-saving upgrades increase the comfort, convenience, and aesthetics of your home.

■ Operating Cost

When you buy an appliance, you pay more than just the sales price — you commit yourself to paying the cost of running the appliance for as long as you own it. These energy costs can add up quickly. For example, running a refrigerator 15–20 years costs as much as the initial purchase price of the unit. That 100-watt light bulb you just put in will cost about $7 in electricity over its (short) life.

The sum of the purchase price and the energy cost of running an appliance or light bulb over its lifetime is called its life-cycle cost. The life-cycle costs of energy-efficient appliances are lower than those of average models even though the latter may cost less to buy. To determine a basic life-cycle cost, use the following equation:

LCC = Initial Cost + (Annual Operating Cost x Years of Operation),

where the operating cost can include energy costs, maintenance and repair. For "years," you would use the expected life of the equipment in question.

■ Rebates

To increase the economic benefits of buying more energy-efficient appliances and boosting your overall home efficiency, check for rebates offered by your local energy and water utilities or tax incentives available from your state or the federal government. Rebates are most common for high-efficiency refrigerators, clothes washers, lighting products, and cooling equipment. Rebate programs are much more common among electric companies than gas companies, although some gas utilities offer rebates for high-efficiency furnaces and boilers. If you plan to buy a major appliance soon, ask your utility if it offers rebates for efficient models.

For More Information:

ENERGY STAR offers a "Rebate Finder" on their website where you can look up whether special offers are available in your area.
www.energystar.gov

For tax credit information, contact your state energy office and refer to the Tax Incentives Assistance Project (TIAP) website.
www.energytaxincentives.org

Energy Use and the Environment

Every time you buy a home appliance, tune up your heating system, or replace a burned-out light bulb, you're making a decision that affects the environment. You are probably already aware that most of our biggest environmental problems are directly associated with energy production and use: global warming, urban smog, oil spills, acid rain, and mercury deposition, to mention a few. You also probably know that driving your car less is one of the best ways to reduce your environmental impact. But you may not realize just how big a difference each of us can make by taking energy use into account in our household purchasing and maintenance decisions.

For example, did you know that every kilowatt-hour (kWh) of electricity you avoid using saves over one and a half pounds of carbon dioxide (CO_2) that would otherwise be pumped into the atmosphere? If you replace a typical 1987, 20-cubic-foot refrigerator with an energy-efficient 2007 model, you'll save more than 500 kWh and almost 1,000 pounds of CO_2 emissions per year!

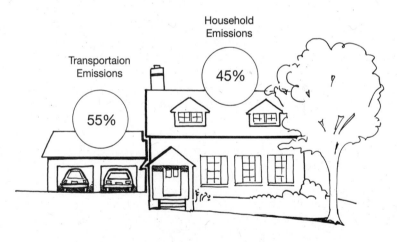

For a typical two-car, single-family household, energy used in the home accounts for almost half of that family's total greenhouse gas contributions and energy costs!

TABLE 1.1 Energy Conservation and CO_2 Savings in the Home

Energy Conservation Measure	CO_2 Savings (tons/yr) (1)			
	Gas	Oil	Electric	Gasoline
Replacing 10 75-watt incandescent light bulbs with 23-watt compact fluorescents (2)	–	–	0.7	–
Replacing typical 1987 refrigerator with energy-efficient 2007 model (3)	–	–	0.4	–
Replacing a 65% efficient furnace or boiler with 90% efficient model (4)	2.0	3.0	–	–
Replacing single-glazed windows with "superwindows" (5)	1.0	1.4	3.7	–
Planting shade trees around house and painting house a lighter color (6)	–	–	0.9-2.4	–
Installing a solar water heating system (7)	0.8	1.4	4.9	–
Boosting energy efficiency of new house under construction from standard levels to super-insulated levels (8)	5.3	7.4	19.8	–
Eliminating two car trips per week (9)	–	–	–	7.8
Replacing average vehicle with hybrid (10)	–	–	–	26.6

Notes:

1 See Table 1.2 for CO_2 emissions factors for each fuel. For gasoline, assumes 157 lbs CO_2/gallon.
2 Assumes lights on five hours per day.
3 Average 1987 model uses 1,000 kWh per year; 2007 model uses 450 kWh per year.
4 Assumes 1,850 ft² house with 6.95 Btu/ft² x °F-day and northern climate (6,300 heating degree-days)
5 Assumes 350 ft² window area; replacing double-glazed, aluminum-framed windows with triple-glazed, dual-low-e, argon-filled superwindows
6 Data from Lawrence Berkeley National Lab, Berkeley, Calif. Based on computer simulations for various locations around the country.
7 Assumes two-panel system providing 14.25 million Btu/year (75% of demand)
8 Boosting energy performance from 6.95 Btu/ft² x °F-day to 1.37 Btu/ft² x °F-day; assumes high-efficiency heating system.
9 Carpooling, biking, or using public transit to eliminate two 20-mile roundtrip commutes per week; assumes vehicle getting U.S. average light-duty fuel economy (2005) of 21.0 mpg.
10 Replacement of average 2005 model vehicle (21.0 mpg) with hybrid getting 40 mpg; assumes vehicle driven 15,000 miles per year.

CO_2 is the number one contributor to global warming, a process that scientists say could raise the earth's temperatures by 3–7°F over the next hundred years. Worldwide, we pump some 27.5 billion metric tons of CO_2 into the atmosphere each year — more than four tons for every man, woman, and child on earth. The United States is responsible for more than 20% of that, or close to 6 billion tons per year. On a per-capita basis, that comes to almost 20 tons for each American, though some of us produce a lot more than others. Reducing CO_2 emissions by a few tons per year may not seem like a lot, but the collective actions of many will have a dramatic effect.

There are numerous energy-saving products and improvements around the home that can help the environment. Table 1.1 shows the reductions in CO_2 emissions achieved from a few energy improvements. With some of these you'll notice different CO_2 savings depending on the type of fuel used. That's because some fuels give off less CO_2 than others.

If you are interested in reducing your carbon footprint, Table 1.2 provides a comparison of the CO_2 emissions from common household energy sources. With this information, it's easy to calculate just how

TABLE 1.2 CO_2 Emissions from Different Energy Sources

Fuel Type	CO_2 produced per unit of fuel	Lbs CO_2 per million btu
Fuel Oil	26.4 lbs/gallon	190
Natural Gas	12.1 lbs/therm	118
Propane	12.7 lbs/gallon	139
Gasoline	23.8 lbs/gallon	190
Wood (1)	2.59 tons/cord	216
Coal (direct combustion)	2.48 tons/ton	210
Electricity from coal only	2.37 lbs/kWh	694
Electricity from oil only	2.14 lbs/kWh	628
Electricity from natural gas only	1.32 lbs/kWh	388
Electricity (national weighted average including all generation types)	1.57 lbs/kWh	460

1 If the wood is harvested on a sustainable basis, there is no net CO_2 emission because the growing trees absorb more CO_2 than is released when burning the wood.

much CO_2 you are introducing into the atmosphere through your energy use. Simply look at your energy bills to find out how much fuel you are using: gallons of oil, therms of natural gas, kilowatt-hours of electricity, etc. Multiply that value by the quantity of CO_2 produced per unit of fuel in Table 1.2.

Carbon dioxide is only one of the environmentally harmful gases resulting from energy use. Others, such as sulfur dioxide, nitrous oxide, carbon monoxide, and ozone, have much more direct effects — effects that can be seen and smelled in every major urban area of the country.

You may notice that CO_2 emissions per unit of energy are much higher for electricity. That difference stems from inefficiencies in the process of converting fuel to electricity and distributing the power through the grid to end-users in their homes and workplaces. Electricity often travels great distances from the power plant to the buildings where it is used. The figure below illustrates the losses attributed to each stage of electricity generation, transmission, and distribution.

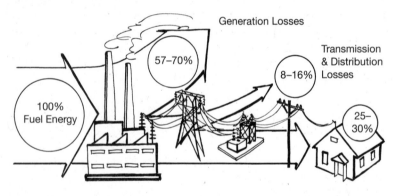

Only one third of fuel source energy reaches your home as electricity.

Despite this drawback, electricity remains vital to our way of life and our economy, and it offers a number of benefits over other fuels for many end-uses. To minimize the negative impacts, we must learn to get the most out of every kWh by using energy as efficiently as possible and looking for new opportunities to support renewable power sources and on-site or local power production.

For More Information:

If you are interested in becoming "carbon neutral," the following resources will help you calculate your carbon footprint and find the most trusted carbon offset companies.

Clean Air, Cool Planet www.cleanair-coolplanet.org

Tufts Climate Initiative www.tufts.edu/tie/tci

The federal government and many state governments have recognized the importance of energy efficiency to our nation's security and economic prosperity. Appliance efficiency standards that took effect in the early 1990s saved more than 88 billion kWh in 2000 — about 28 million tons of CO_2. Updates to these standards will save more than 250 billon kWh in 2010. Despite these impressive gains, standards only eliminate the lowest efficiency products from the market. It is up to consumers to do the rest and demand more from the marketplace. If the roughly 40 million households in climates with large heating needs boosted their furnace or boiler efficiencies from 70% to 90%, some 45 million tons of CO_2 emissions would be eliminated each year. Substituting compact fluorescent lamps for the ten most frequently used incandescent lamps in every house in the country would reduce CO_2 emissions by about the same amount!

To get a sense of just how effective energy conservation can be, take a look at the 1970s and 1980s. From 1973 to 1986, the U.S. gross national product grew 36% with no increase in energy use at all. Had efficiencies remained at 1973 levels, we would be spending an extra $150 billion in energy bills each year and pumping 1½ times more CO_2 into the atmosphere! We are already saving the equivalent of 13 million barrels of oil each day — half of the OPEC output — and, compared with 1973 projections, we're getting by with 250 fewer large power plants than would have otherwise been required.

Planning Energy Improvements

As you think about how to reduce your environmental impact and energy bills, it can be hard to know where the best opportunities lie. The "Home Energy Checklist for Action" preceding the start of this chapter (page xii) provides one way to prioritize some common home improvements. It is also useful to understand how your appliances stack up in terms of energy use, as shown in the pie chart on the next page. Keep in mind that, although heating and cooling consume by far the most energy, your best opportunities for reducing these pieces of the pie may not come from replacing equipment but from improving the efficiency of the building itself.

HOW TO USE THIS BOOK

The *Consumer Guide to Home Energy Savings* is the most complete and up-to-date guide available on energy savings in the home. Following a review of measures to tighten up the building shell itself, the book focuses on the things you put in it your home—including major appliances, heating equipment, air conditioning, lighting, and electronics—and how the energy use of those products can be reduced.

If you're about to buy a new appliance or heating system, you'll be most interested in the tips on what to look for when buying new equipment. Otherwise, look for guidance on how to get the best energy performance through operation and maintenance of the products you already own. For further information and updates, we've included links directing you to valuable online resources provided by ACEEE and many others.

The break-down of energy use in a typical home

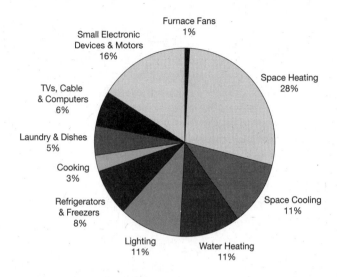

Furnace Fans
1%

Small Electronic
Devices & Motors
16%

Space Heating
28%

TVs, Cable
& Computers
6%

Laundry & Dishes
5%

Cooking
3%

Refrigerators
& Freezers
8%

Space Cooling
11%

Lighting
11%

Water Heating
11%

Source: U.S. Department of Energy, 2007.
www.eia.doe.gov

■ Setting Priorities

The decision to make certain energy improvements can be obvious — if you have a broken appliance and need to replace it, for example, use this book to make a smart purchase decision. But there may be other important priorities for your house that you are unaware of. If this is the case, you may want to perform a quick self-audit, or go ahead and hire a professional to help find the most cost-effective improvements (see Chapter 2).

Some of the more involved energy improvements mentioned here, such as replacing windows and insulating, make the most sense when you are planning other remodeling work. If you are going to extend a wall out to enlarge your kitchen or put in a larger dormer for a master bedroom expansion, by all means boost energy efficiency at the same time. Re-build walls with high insulation levels. Put in high-performance insulating windows.

For More Information:

Enter your zip code and some details about your home into the online Home Energy Saver tool to see what energy savings are possible based on the current conditions of your house.
www.hes.lbl.gov/hes

As long as you're ripping out walls, take advantage of the mess and go a little further, boosting the efficiency of some of the adjoining walls and windows as well. With a small addition, some of this work might even pay for itself right away if it means, for example, that you can get by without adding a separate heating system or expanding your current heat distribution system.

■ Understanding ENERGY STAR

Once you've identified your high-priority areas and are ready to look for new products, look for the ENERGY STAR. The U.S. Environmental Protection Agency (EPA) and U.S. Department of Energy (DOE) recognize the local and global environmental significance of energy-efficient products. Working in voluntary cooperation with manufacturers and retailers, these agencies have created a distinctive ENERGY STAR label to help consumers identify energy-efficient heating and cooling equipment, appliances, computers,

ENERGY STAR qualifies over 50 different types of products.

lighting, and home electronics. Some homebuilders are now constructing entire ENERGY STAR homes, which include a variety of energy-efficient features and equipment. ENERGY STAR homes are at least 30% more energy efficient than the current International Energy Conservation Code (formerly known as the Model Energy Code). You can even get an ENERGY STAR-qualified whole-house retrofit to optimize your overall home energy performance.

■ Understanding EnergyGuide Labels

Federal law requires that EnergyGuide labels be placed on all new refrigerators, freezers, water heaters, dishwashers, clothes washers, room air conditioners, central air conditioners, heat pumps, and furnaces and boilers. These labels are bright yellow with black lettering. The EnergyGuide label can be useful when evaluating how the specific product you are considering compares to other products of the same type. EnergyGuide labels are not required on kitchen ranges, microwave ovens, clothes dryers, portable space heaters, and lights. For these products, look for the energy-conserving features discussed throughout this book.

The main feature of the label is a line scale showing how that particular model compares in energy efficiency with other models on the market of comparable size and type. You will see a range of lowest to highest. A word of caution — the ranges shown on the labels are not updated frequently, and manufacturers are constantly introducing new appliances.

For refrigerators, freezers, water heaters, dishwashers, and clothes washers, the range shows annual energy consumption (in kWh/year for electricity or therms/year for gas). The most efficient models will have labels showing energy consumption (represented by the downward-pointing triangle labeled "This Model Uses"), at or near the left-hand end of the range. It will be close to the words "Uses Least Energy." An estimate of annual operating cost is also provided.

For room air conditioners, central air conditioners, heat pumps, and furnaces and boilers, the range is not energy consumption, but rather the energy efficiency ratings for these products (EER, SEER, HSPF & SEER, and AFUE, respectively). Therefore, labels on the most efficient models will show "This Model's Efficiency" at or near the right-hand end of the range, close to the words "Most Efficient." To estimate your operating costs for these products, refer to the manufacturer's fact sheets available from the seller or installer.

For More Information:

In February 2008, a revised label will appear on all appliances. To learn more, go to www.ftc.gov/appliances

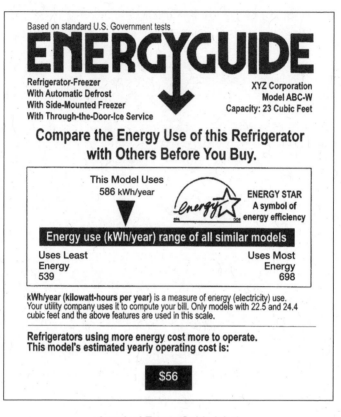

A typical EnergyGuide label

■ Quality, Reliability and Availability

When shopping for major home appliances, you may want to call several stores or dealers to check the price and availability of different models that you find in the ENERGY STAR or other listing. You can ask the salesperson for information about the efficiency of each model, but be aware that he or she may not know very much about energy performance. Take this guide along when you shop to make sure the appliances contain recommended energy-efficient features. Many retailers are beginning to include more energy use information on the products they carry on their websites.

Also keep in mind that energy performance is not the only consideration you should use when selecting home appliances. Consumers must consider how effectively the appliances perform their primary functions — cleaning dishes, keeping food cold, etc. — as well as how much energy they use in doing it. For example, you wouldn't consider buying a dishwasher that didn't get your dishes clean, even if it used just half as much energy. This book does not pretend to be a comprehensive review of product reliability or performance, or a guide to convenience features found in these products; there are other sources for that information (See the Appendix). It is worth noting, however, that energy-efficient appliances are generally high-quality products due to the better materials and components used in their construction.

For the latest innovative energy-saving technologies, consider doing some additional research using the growing list of newsletters and journals (both online and print) that cover green products. Some top publications are listed in the Appendix under "Home Energy Publications and Newsletters."

New Homes

If you are building a new house or a major addition, do it right the first time, saving money and the environment for decades to come. Today's state-of-the-art energy-efficient houses typically require less than a quarter as much energy for heating and cooling as most existing houses. There are thousands of homes in the northern United States and Canada with yearly energy bills that total just $200 to $300.

These homes cost more to build than a standard house, but not that much more. You might spend an extra $5,000–$10,000 to build a super-efficient house with R-30 walls, R-38 ceilings, R-19 foundations, R-3 windows, and very low air leakage. But that extra cost will usually be recovered in just five to ten years through energy savings. Plus, you'll be more comfortable and you'll have the satisfaction of knowing that your house is dumping less pollution and carbon dioxide into the atmosphere.

Once you get an idea of what you want, contact builders or architects in your area and find out how experienced they are with energy-efficient construction. Special skills are required to build high-efficiency houses and to install features such as heat-recovery ventilation systems. You may need to spend a little extra time looking for the right builder, but the time and effort will be well worth it.

Sustainability

Saving energy is one of the greatest impacts you can have on reducing environmental degradation. But it does not make your lifestyle environmentally sustainable. The *Consumer Guide to Home Energy Savings* touches on some important considerations like water conservation, indoor pollutants and appliance recycling, but the book is not exhaustive. If you are interested in becoming more environmentally sustainable, it is important to consider the materials and chemicals that comprise your products as well as the resources that go into their manufacturing, transport, and disposal. This applies not just to household equipment but to all items you purchase. Look for ways to reduce your overall consumption. Buy locally grown foods. Reduce the number of miles you drive by combining trips, carpooling, biking, or taking public transportation, when feasible. By choosing a more sustainable lifestyle, we contribute to making a safer, healthier environment for ourselves and our children. And you just might find these changes improve your overall quality of life in surprising ways.

For More Information:

To learn more ways to improve your environmental sustainability, refer to resources in the Appendix of this book or go to www.aceee.org/consumerguide/resources.html.

Chapter 2

The Building Envelope

If you live in a cold climate, you probably spend something like half of your energy dollars on heat. Your old furnace or boiler chugs away burning gas or oil like there's no tomorrow. So should you rush right out and buy a new super-efficient one? Not necessarily.

Replacing your existing heating system with one that's more efficient may well be a wise step, but it shouldn't necessarily be your first step. You should first try to lower your heating requirements. Tighten up. Weatherize. Insulate. By reducing your heating needs, you may increase comfort and you may be able to get by with a significantly smaller — and less expensive — furnace or boiler.

The same arguments hold true with air conditioning. If you live in a warm climate with high cooling requirements, it makes a lot of sense to tighten up the house to reduce your cooling load before investing in new air conditioning equipment.

A tight, well-insulated house saves energy and allows you to get by with smaller-capacity heating and air conditioning systems, and it is also more comfortable, with smaller temperature swings. No more cold drafts at your feet while temperatures at head level are a sweaty 80°F. With less of this temperature stratification during the winter months, you'll even find yourself comfortable at a lower thermostat setting than you're used to. In the next few pages, we'll take a look at some of the measures you can take to improve the energy efficiency of your house, and when it makes sense to consider such projects.

Home Energy Auditors, Raters and Contractors

Before making major efficiency improvements to your house, find out from a pro where and why energy is being wasted and what you should do about it. Today's home energy specialists — sometimes called house doctors, energy auditors, raters, or home performance contractors — go beyond simple checklist audits. They study the building as a system, performing full check-ups that are designed to address overall safety, comfort, energy efficiency, and indoor air quality. After all, many construction flaws can result in high energy bills, and some conditions that cause high energy bills can compromise building safety or resident health and comfort (more about that in Chapter 3). For example, home performance contractors can diagnose and fix uneven heat distribution while lowering energy costs, or reduce cold drafts and moisture condensation problems while reducing energy bills.

These contractors use sophisticated equipment, like blower doors and infrared cameras, to help pinpoint air leaks and areas with inadequate insulation. Auditors that are properly certified should also test and tune-up your heating and cooling equipment, and check for duct leakage (see Chapters 4 and 5). These professionals know what to look for in both newer and older houses, and the investment in their time is usually well worth the cost. They will often perform the air sealing work as they go, and can usually connect you with qualified contractors to complete major work or work with your own contractor to get the right work done. Some utility companies provide basic energy audits free of charge.

A home energy rating entails the same kind of on-site diagnostic tests that an energy auditor would do; but with a rating, your house will be given a point score between 1 and 100 that compares your house to others. Energy ratings are used more often for new homes to obtain an ENERGY STAR distinction or energy mortgage, but the tool can be useful for existing homes as well. A rater will look at your energy bills to complete a full analysis of your home's energy costs and improvement options, and the score acts as a tool to evaluate and pinpoint the most cost-effective improvements. Find a rater that is certified by RESNET.

Common Home Performance Diagnostic Tests

A **blower door** is used to determine your home's airtightness. It is a powerful fan that mounts into the frame of an exterior door and pulls air out of the house in order to lower the inside air pressure. While operating, the auditor can determine infiltration rate and better identify specific leaks around the house. Make sure your auditor is using a "calibrated" blower door.

A **duct blaster** is sort of like a blower door for your duct system, except instead of pulling air out, it uses a fan to force air into the system until, it reaches a standard pressure. The fan is attached to one entry into the duct system while all other registers are temporarily sealed off. While under pressure, the instrument can detect how much air is leaking from the duct work. See Chapter 4 for more on duct sealing.

Using a **digital infrared camera** is a very powerful way to detect both thermal defects and air leakage. Infrared cameras measure surface temperatures — the camera "sees" variations in heat instead of light, and expresses warmer areas with warmer colors. So you can actually see if, for example, a portion of your wall is missing insulation, without having to drill a hole. Interpreting the images correctly, however, depends on experience and understanding the indoor and outdoor conditions.

In select areas, there may be home performance contractors trained under the Home Performance with ENERGY STAR program. This is a U.S. EPA program that partners with local states, utilities and municipalities to train a network of contractors to conduct home energy improvements in existing homes. These contractors act as general managers that oversee the whole process from testing through installation of new equipment. Any improvements purchased may be partially subsidized by the local agency up to a certain dollar amount or qualify for low-interest financing.

If you are a renter, encourage your landlord to have an energy audit and follow through on the recommended energy improvements. You might offer to help your landlord by arranging for the audit and even doing some of the work in exchange for rent. After all, if you pay for heating, air conditioning, and electricity, energy improvements are very much in your interest. Even if your landlord won't pay for energy conservation projects, many of the suggestions in this chapter are inexpensive

**An energy auditor uses sophisticated equipment
to test where heat enters or escapes your house.**

For More Information:

Learn more about contractor services, home improvement
incentives in your state, and local contractors
certified by the Building Performance Institute.
www.bpi.org ◼ www.hometuneup.com

Or try finding an energy rater through RESNET. These contractors
should also be able to discuss energy-efficient mortgages.
www.resnet.us/directory ◼ (760)806-3448

Find out whether Home Performance with ENERGY STAR
is available in your area, on the ENERGY STAR
website under "home improvement."
www.energystar.gov ◼ (888) 782-7937

enough that they'll pay for themselves in just a year or two, justifying your out-of-pocket expenditures.

Depending on your income, you may qualify for a free energy audit and energy efficiency improvements to your home through the Weatherization Assistance Program.

For More Information:

To contact your local weatherization agency visit this website and click on "How Do I Apply for Weatherization."

www.eere.energy.gov/weatherization

Find and Seal Air Leaks

Hidden air leaks cause some of the largest heat losses in older homes. In the average home, small openings in the outer shell of a house account for almost 30% of total heat lost. Common air leakage sites are listed below:

- Plumbing penetrations through insulated floors and ceilings
- Chimney penetrations through insulated ceilings and exterior walls
- Along the sill plate and band joist at the top of foundation walls
- Fireplace dampers
- Attic access hatches
- The tops of interior partition walls where they intersect with the attic space
- Recessed lights and fans in insulated ceilings
- Wiring penetrations through insulated floors, ceilings, and walls
- Missing plaster
- Electrical outlets and switches, especially on exterior walls
- Window, door, and baseboard moldings
- Dropped ceilings above bathtubs and cabinets
- Kneewalls in finished attics, especially at access doors and built-in cabinets and bureaus.

The appropriate material for sealing these hidden air leaks depends on the size of the gaps and where they are located. Caulk is best for cracks and gaps less than about 1/4" wide. In choosing caulks, read the label carefully to make sure that the caulk is suitable for the material to be sealed. Look for caulks that remain flexible over a 20-year lifetime. If the caulked joint will be visible, choose a paintable caulk or one that is the right color. In general, you should avoid the cheapest caulks, because they probably won't hold up well. Expanding foam sealant is an excellent material to use for sealing larger cracks and holes that are protected from sunlight and moisture. One-part polyurethane foam is commonly available in hardware and building supply stores. Today's products are safe for atmospheric ozone.

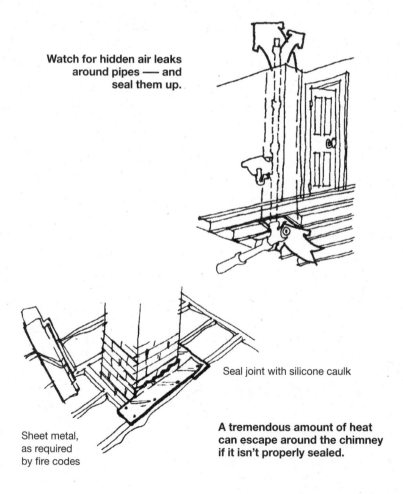

Watch for hidden air leaks around pipes — and seal them up.

Seal joint with silicone caulk

Sheet metal, as required by fire codes

A tremendous amount of heat can escape around the chimney if it isn't properly sealed.

Backer rod or crack filler is a flexible foam material, usually round in cross-section (1/4" to 1" in diameter), and sold in long coils. Use it for sealing large cracks and to provide a backing in very deep cracks that are to be sealed with caulk.

Use rigid foam insulation for sealing very large openings such as plumbing chases and attic hatch covers. Fiberglass insulation can also be used for sealing large holes, but it will work better if wrapped in plastic or stuffed in plastic bags, because air can leak through exposed fiber-

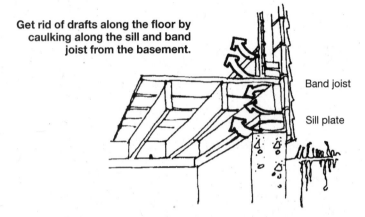

Get rid of drafts along the floor by caulking along the sill and band joist from the basement.

Band joist

Sill plate

Dropped ceilings above closets, showers, and cabinets can be among the worst offenders when it comes to air leakage.

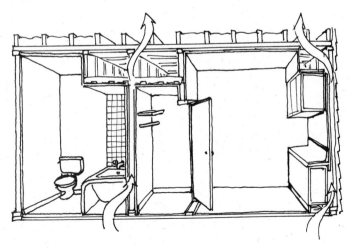

glass. Don't use plastic in places where high temperatures may be reached, and always wear gloves and a dust mask when working with fiberglass.

Sheets of polyethylene can be taped over large holes to block airflow in some situations, but this is usually a fairly temporary measure, since the polyethylene may disintegrate over time if not protected. Specialized materials such as metal flashing and high-temperature silicone sealants may be required for sealing around chimneys and flue pipes. Check with your building inspector or fire marshal if unsure about fire-safe details in these locations.

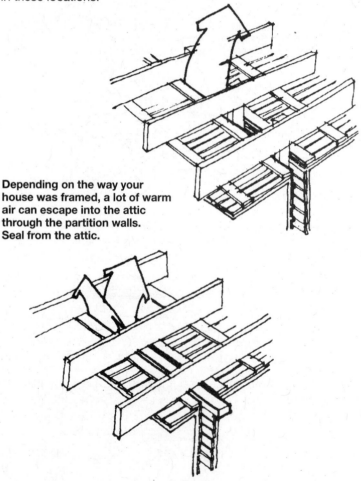

Depending on the way your house was framed, a lot of warm air can escape into the attic through the partition walls. Seal from the attic.

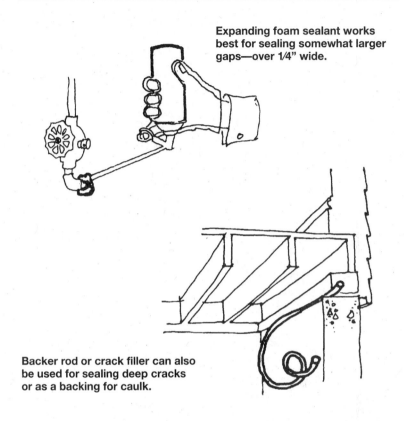

Expanding foam sealant works best for sealing somewhat larger gaps—over 1/4" wide.

Backer rod or crack filler can also be used for sealing deep cracks or as a backing for caulk.

Insulate

Insulation is your primary defense against heat loss through the house envelope. Heat losses through the wall, roof and floor together amount to over 45% of all the heat a typical house loses. In the summer, heat entering through the roof and walls in particular account for one fourth of your cooling needs. See Table 2.1 for recommended insulation levels for your climate.

Putting insulation into a house after it is built can be pretty difficult. Because of the large area involved, walls are most important. You may be able to tell if the walls are insulated by removing an outlet cover and peering into the wall cavity. Another way to check for insulation is to find a closet (or cabinet) along an exterior wall. Drill two 1/4" holes into the wall about 4" apart, with one hole above the other. Shine a flashlight into one hole while looking into the other, and any insulation should be apparent. If there isn't any insulation, the best option is to bring in an insulation contractor to blow cellulose or fiberglass into the walls.

TABLE 2.1 Cost-Effective Insulation R-Values for Existing Homes

Climate	Heating System	Ceiling	Wood-frame Wall	Floor	Basement or Crawl Space Walls
WARM high cooling and low heating needs	gas/oil or heat pump	R-22 to R-38	R-11 to R-13	R-11 to R-13	R-11 to R-19
	electric resistance	R-38 to R-49	R-13 to R-25	R-13 to R-19	R-11 to R-19
MIXED moderate heating and cooling needs	gas/oil or heat pump	R-38	R-11 to R-22	R-11 to R-13	R-11 to R-19
	electric resistance	R-49	R-11 to R-26	R-13 to R-19	R-11 to R-19
COLD high heating and low-moderate cooling needs	gas/oil or heat pump	R-38 to R49	R-11 to R-22	R-25	R-11 to R-19
	electric resistance	R-49	R-11 to R-28	R-25	R-11 to R-19

Notes:
a. Insulation is also effective at reducing cooling bills. These levels assume your house has electric air-conditioning.
b. For wood-framed walls, R-values may be achieved through a combination of cavity insulation and rigid board insulation and are for insulation only (not whole wall).
c. Do not insulate crawl space walls if crawl space is wet or ventilated with outdoor air.

Source: Adapted from the U.S. Department of Energy, 1997, Insulation Fact Sheet.

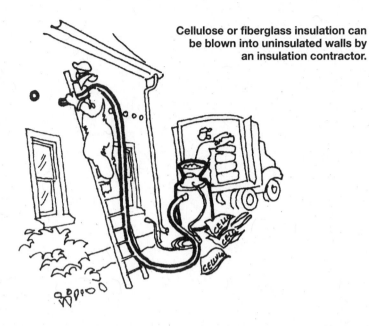

Cellulose or fiberglass insulation can be blown into uninsulated walls by an insulation contractor.

Adding insulation to an unheated attic is usually a lot easier. If there is no floor in the attic, simply add more insulation, either loose fill or unfaced fiberglass batts. If the existing insulation comes up to the top of the joists, add an additional layer of unfaced batts across the joists. This helps to cover gaps between the first layer of batts. If an attic floor is in place, you may need to remove that floor before adding insulation (be very careful not to step through the ceiling below!). In most of the country, a full foot of fiberglass or cellulose insulation is cost-effective in the attic floor.

If the attic is finished with a sloped cathedral ceiling, adding extra insulation is much more difficult. If there is no insulation there at present, it may be worth pulling off the drywall, insulating, and installing a new ceiling. Or, if you will be re-roofing in the near future, consider adding a layer of rigid foam insulation and decking on top of the existing roof and then shingling over that. In either case, it's a major project, and you will probably have to bring in a contractor to do the work. If there is already some insulation in the roof, it may be hard to justify the cost of this work, in which case you can focus your energy efforts elsewhere.

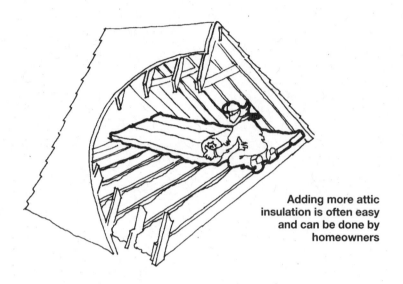

Adding more attic insulation is often easy and can be done by homeowners

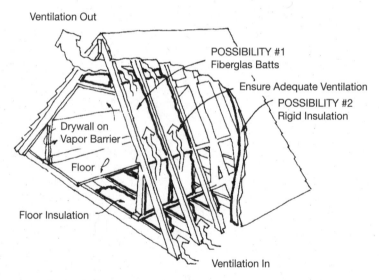

Ventilation Out

POSSIBILITY #1
Fiberglas Batts

Ensure Adequate Ventilation

POSSIBILITY #2
Rigid Insulation

Drywall on
Vapor Barrier

Floor

Floor Insulation

Ventilation In

Insulating a sloped ceiling in a finished attic can be done from the inside, or by installing foam and new roofing on the outside.

In the basement, heat loss through foundation walls is often neglected even in new houses. But in fact, in an otherwise well-insulated and tight house, as much as 20% of the total heat loss can occur through uninsulated foundation walls. Insulating the foundation or floor can easily save several hundred gallons of oil or several hundred therms of gas per year in northern climates.

Increasingly, building codes require basement insulation, and it is essential for basements used as living spaces in almost all climates. However, controlling moisture in these below ground spaces is a real challenge and complicates the insulation process. Moisture-related failures of basement insulation can give rise to mold, mildew and other problems.

Materials that could be damaged by moisture, such as fiberglass batts and cellulose, should never be used to insulate a basement. Interior vapor barriers can also be very damaging because they prevent basements from drying to the inside. Interior basement insulation should start with rigid foam installed against the basement walls.

If you have a crawl space, it should be sealed rather than ventilated, in most climates. To do this, use 6-mm thick polyethylene sheeting as a moisture barrier to cover the ground and seal tightly to walls and columns. Then use rigid foam to insulate the foundation walls. In the South, it is important to keep an uninsulated band for inspection of possible termite tunnels.

If you are considering finishing your basement and using it as a living space, seek the advice of an experienced professional. Because of the challenges inherent in constructing basement wall systems that can provide healthy and comfortable conditions, it may make more sense to build an above-grade addition to your house, rather than trying to make the basement liveable.

Upgrade Inefficient Windows and Doors

About one-quarter of a home's total heat loss during colder months usually occurs through windows and doors. In the summer, about one third of the heat that enters a house is from the sun shining through. But windows are not a total negative in terms of your home's energy performance. On the positive side, windows can be used for passive solar heating in the winter months to help reduce heating costs, and of course windows provide natural daylighting and views to the outside world.

Because windows outnumber doors, we devote the vast majority of this section to them. That said, some of the considerations about windows apply to doors as well, like the importance of weatherstripping or frame material.

■ Understanding How Windows Work

To understand how windows affect heating and cooling costs, we need to know a little about how energy flows through them. The next illustration shows the primary mechanisms of heat transfer through windows.

Sunlight. Solar radiation is an important source of heat, and is transmitted directly through most windows. Solar radiation consists of visible light and a part of the solar spectrum that is heat but not visible light (infrared heat radiation). Glazings are available that selectively transmit light of different wavelengths—for example, some appear totally clear (high visible light transmittance) while blocking infrared heat radiation. A window's solar heat gain coefficient (SHGC) is the measure of the amount of solar energy that passes through the window; typical values range from 0.4 to 0.9, and the higher the SHGC, the greater the percentage of solar energy that is transmitted to the inside.

How heat moves through windows.

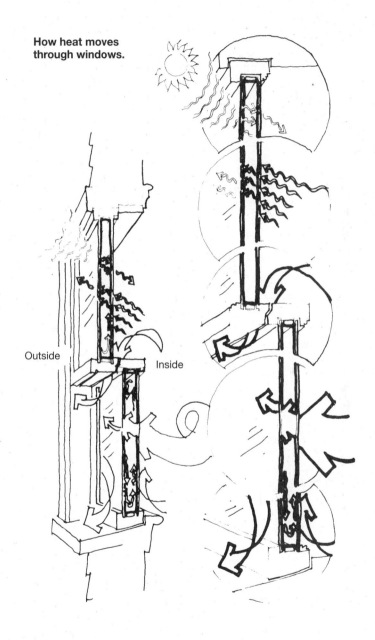

Outside

Inside

Radiation. Radiant heat is given off by warmer objects to colder objects. Things warmed by sunlight become stronger sources of radiant heat, and radiant heat is blocked by most window glazings. Greenhouses and passive solar heating illustrate the dynamic between sunlight and radiant heat: sunlight passes through glazings, warming objects indoors, but the heat from those objects does not quickly escape back through the glazings, and the space warms up. Special coatings on high-performance windows change the way heat radiation is absorbed and reradiated (see discussion on low-e glazings below).

Conduction. Conduction is the mechanism of heat transfer through physical contact. Heat conducts from the warmer to the cooler side of a window as each molecule excites its neighbor, passing the energy along. Conduction occurs not only through solid materials (window glass and frames), but also through the air space between the layers of glass.

Convection. Convection is the movement of heat in a medium, such as air. The heat is transferred as the molecules of air are physically moved from one place to another. A warm glass surface heats the air next to it, causing the air to rise. A cold glass surface is warmed by the air next to it, and that air mass will fall as it gives up its heat. These convection currents occur on the inside of a window, on the outside, and even between layers of glass if they are separated too far from each other.

Air leakage. Infiltration is the process that carries heat through cracks and gaps around window frames. Infiltration through leaky windows can carry cold air into a house and warm air out. Infiltration is driven by wind and other differences in air pressure, such as warm air rising inside a house.

Outside

Inside

A tremendous amount of cold air can leak in through old windows.

■ Should I Replace?

To reduce heat loss and heat gain, you can either fix up your existing windows or replace them with new energy-efficient units. Many people who are worried about high heating costs turn to new windows as an early upgrade. Although new windows are great, they can be expensive, and they rarely pay back their cost quickly enough to be a good investment on energy savings alone. It's likely that your money would be much better spent on a top-notch energy audit and repair of your insulation, thermal envelope, and ductwork. However, if you plan to replace windows anyway—for comfort, appearance, cleaning convenience, modernization, or any other reason—it almost always pays to invest the small added cost in highly efficient windows rather than minimum-performance ones.

If your existing windows have rotted or damaged wood, cracked glass, missing putty, poorly fitting sashes, or locks that don't work, you may well be better off replacing them. If the windows are generally in good shape, it will probably be more cost-effective to boost their efficiency by weatherstripping, caulking, and fitting them with storm panels as described in the following section.

■ Fixing Up Existing Windows

Unless your windows are visibly damaged, there are several simple repairs, from cheap and temporary to more permanent, that will pay for themselves by lowering your heating and cooling bills.

Rope caulk. The quickest and least expensive option is to seal all window edges and cracks with rope caulk. This costs less than $1 per window and only takes a few minutes. Rope caulk may be taken off, stored in foil, and reused for two or three seasons, but once it hardens you should discard it. The upper sash of double-hung windows can be permanently caulked closed if you don't need to open it.

Weatherstripping. A more permanent solution is to weatherstrip the windows. This is more time-consuming and expensive than installing rope caulk ($8–$10 per window), but it only needs to be done once, it permits you to open the window, and the weatherstripping is out of sight. The type of weatherstripping to use depends on the type of window — both compression-type and V-strip weatherstripping are widely available in home improvement stores. You usually get what you pay for with weatherstripping products, so spend the few extra dollars necessary to get a top-quality product that is likely to hold up over years of use.

While you're weatherstripping, don't forget about your doors. As with windows, make sure your doors are in good shape. Weatherstrip around the whole perimeter to ensure a tight seal when closed. Install quality door sweeps on the bottom of doors if they aren't already in place. On an old, uninsulated metal or fiberglass door, a storm door probably isn't cost-effective. In fact, a glass storm door may damage the plastic trim on some metal or fiberglass doors by trapping heat.

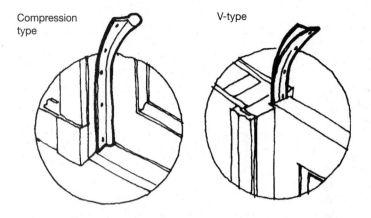

Compression type

V-type

Weatherstripping is the most permanent way to cut air leakage through windows and doors.

Install door sweeps to reduce airflow underneath doors.

Storm windows. The next step in improving window energy efficiency is to install some type of storm window. If you have single-glazed windows, storm panels will double their energy efficiency. If you heat with oil, the improved energy efficiency in a cold climate will save about a gallon of oil per square-foot of window per year. (That's an awful lot of oil that is otherwise leaking out of your house and adding to our air pollution woes.) With gas heat, the savings can top 1 therm per square foot of window per year.

The simplest type of storm window is a plastic film taped to the inside of the window frame. Inner storm window kits are readily available from hardware stores. They cost just $3–$8 per window and typically last for one to three years. Some are made of special shrink-tight plastic that you heat with a blowdryer after installation to get rid of wrinkles. These inexpensive plastic films are especially suitable for apartments and condominiums where exterior improvements are not allowed or are not practical.

Removable or operable storm windows with glass or rigid acrylic panes generally make more sense if you plan to stay in the house for more than a few years. Both exterior and interior storm windows are available, though exterior units are far more common. Most people choose aluminum-framed combination storm/screen windows, which are very convenient to operate. Be careful, though. There are big differences among products on the market, especially relative to air tightness. The tightest units have air leakage rates as low as 0.01 cubic feet per minute (cfm) per foot of window edge. Don't settle for anything higher than 0.3 cfm per foot. Also look for units with low-e coatings on the glass to improve the energy performance.

Storm windows typically cost between $50 and $120, depending on size, quality, and labor for installation — far less than replacement windows. Before buying new storm windows, though, check to make sure you don't already have a stack in a dusty corner of your basement or attic. If these are older wood-framed storm windows, they probably just need a coat of paint. Although they aren't quite as convenient as combination storm/screen windows (they have to be taken down and put up each year and separate screens are required), wooden storm windows are often more energy efficient.

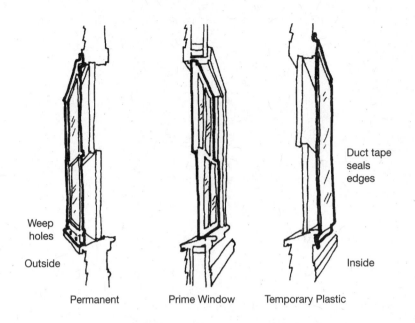

Weep holes

Outside

Duct tape seals edges

Inside

Permanent   Prime Window   Temporary Plastic

Storm windows can be installed on the inside or outside.

If aluminum combination storms are already in place, examine them to make sure they are tightly sealed where mounted to the window casings; if not, caulk all cracks (but do not seal the small weep holes on the bottom edges).

Finally, you can boost the energy efficiency of windows by installing insulating curtains or drapes on the interior. These can be closed at night to significantly cut down on heat loss. They can also be closed on hot summer days to keep out unwanted heat gain. Look for shades or drapes that fit into tracks to keep air from passing around the edges.

■ Buying New Windows

Because of the impact windows have on both heat loss and heat gain, proper selection of products can be confusing. To add to the complexity, window glazing technology has changed tremendously in recent years. The best window glazings today insulate almost four times as well as the best commonly available windows from just fifteen years ago. Because of the rapid pace of change, even skilled designers are often not fully aware of the potential these new glazings offer for energy-efficient building design. The National Fenestration Rating Council (NFRC) is the principle organization responsible for evaluating and labeling windows based on their energy performance. How to interpret their label and the specific performance properties to look for is discussed below. First, let's walk through some of the basic features to look for in a new window.

Reliability and good installation. Choose windows with good warranties against the loss of the air seal. If the glazing seal is lost, not only will fogging occur, but also any low-conductivity gas between the layers of glass will immediately be lost. Consumers should recognize that the manufacturer's quality control at the factory and care during shipping can have a big impact on the window's air tightness at a site. Furthermore, it is senseless to invest thousands of dollars in new windows only to have amateurs install them in your home. The high-rated performance of a window is meaningless if it is installed improperly with gaps and air leaks around the window. Be sure to have experienced contractors install your high-tech windows.

Proper dimensions. To maximize energy performance, choose windows with larger unbroken glazing areas instead of multi-pane or true-divided-light windows. Applied grills that simulate true-divided-light windows are fine; they do not reduce energy efficiency.

Different window geometries.

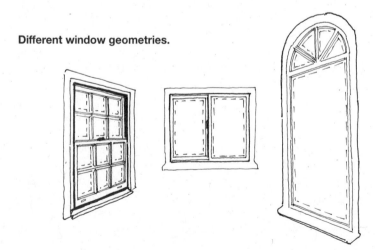

Frame material and sash construction. Window frame and sash construction has a big impact on energy performance. Wood is still the most common material in use, and it insulates reasonably well. Aluminum has been used extensively, particularly in the western part of the U.S., but unless a thermal break is incorporated into the design, aluminum frames conduct heat very rapidly and are therefore inefficient. Vinyl (PVC) windows are gaining in popularity, especially in the replacement market, and some vinyl frames insulated with fiberglass insulate better than wood. "Pultruded" fiberglass frames are entering the residential market from the commercial sector. They insulate well and their cost generally falls between that of wood and vinyl.

Air-tightness. Windows vary dramatically in how effectively they block infiltration. In general, casement and awning windows are tighter than double-hung and other sliding windows. This is because when a casement or awning window is closed, it is pulled in against a compression-type gasket. Sliding windows have to use seals that permit the sash to slide, so they are rarely as airtight. You will find, though, that double-hung windows from a few manufacturers are tighter than casement windows from others, so it makes a lot of sense to examine air leakage specifications carefully when selecting windows.

Multiple layers of glazing. Until the 1980s, the primary way manufacturers improved the energy performance of windows was to add additional layers of glazing. Double glazing insulates almost twice as well as single glazing. Adding a third or fourth layer of glazing results in further improvement. In the 1970s, with rising concern over energy, triple-glazed and even quadruple-glazed windows entered the market. Some of these windows use glass only; others use thin plastic films as the inner glazing layer(s).

Thickness of air space. With double-glazed windows the air space between the panes of glass has a big effect on energy performance. A very thin air space does not insulate as well as a thicker air space because of heat conducted through that small space. During the 1970s a lot of window manufacturers increased the thickness of the air space in double-glazed windows from 1/4" to 1/2" or more. If the air space is too wide, however, convection loops between the layers of glazing occur. Beyond about 1", you do not get any further gain in energy performance with thicker air spaces.

Low-conductivity gas fill. By substituting a denser, lower conductivity gas such as argon for the air in a sealed insulated glass window, heat loss can be reduced significantly. The largest window manufacturer in the country today, Andersen Windows, uses argon gas-fill in all of its insulated glass windows, and most major manufacturers offer argon gas-fill as an option. Other gases that are being used in windows include carbon dioxide, krypton, and argon-krypton mixtures.

Edge spacers. The edge spacer is what holds the panes of glass apart and provides the airtight seal in an insulated glass window. Avoid traditional hollow aluminum spacers because they have extremely high conductivity. Instead, choose edge spacers that are thin-walled steel, silicone foam or butyl rubber. Generally, better edge seals are a low-cost option when ordering windows, and worth considering. The net effect of improved edge spacers can be a 2–10% improvement in window energy performance, depending on the other performance characteristics of the window. With new edge spacers, however, pay particular attention to warranties against seal failure, which results in fogging and loss of any low-conductivity gas-fill.

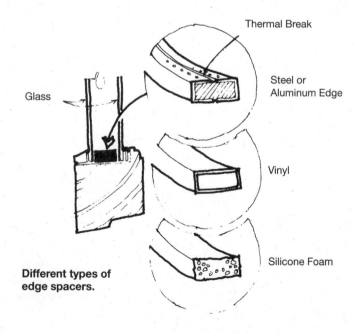

Thermal Break

Glass

Steel or
Aluminum Edge

Vinyl

Silicone Foam

**Different types of
edge spacers.**

Glazing with low emissivity. Tinted glass and tinted window films have long been used in commercial buildings to reduce heat gain through windows. Today, low-e coatings made of a thin, transparent layer of silver or tin oxide are used on high-performance windows to reduce the solar heat gain without reducing visibility as much as older tinted glass. The coatings permit visible light to pass through, but they effectively reflect infrared heat radiation back into the room, reducing heat loss in winter. The variety and placement of the low-e coating on the window varies for different climate zones and applications. Low-e windows with high solar heat gain coefficients are appropriate for northern climates where passive solar heating is advantageous, while "southern low-e" windows with low heat gain coefficients are appropriate in milder climates where summer cooling is more necessary.

Match the application. The different properties of low-e glazings allow you to choose different types of glazing for different sides of your house. For example, if you want to benefit from passive solar heating for the south side, choose windows containing a top-performing low-e glass with a high solar heat gain coefficient. On the north, install the lowest U-value windows you can afford. Or to keep things simpler, you can order the same glazings for the east-, west-, and north-facing windows.

Some window manufacturers now produce both "northern" and "southern" climate low-e products. But other window manufacturers may still offer just one type of low-e glazing as their standard, and charge extra for substitutions—if they provide options at all. So a decision to choose different glazings for windows of different orientations may require some extra shopping around. If you do order different glazings for your different windows, be sure to keep track of which windows have which type of glazing, because they will probably all look identical!

For More Information:

ENERGY STAR offers guidance for the kinds of low-e coatings appropriate for cold, moderate, and warm climates.
Some manufacturers make windows for all climates,
while some only specialize in one.
Search under "Products" for "windows, doors and skylights."
www.energystar.gov

Look for the ENERGY STAR. Windows, doors, and skylights qualifying for the ENERGY STAR label must meet requirements tailored for the country's four broad climate regions: northern, north-central, south-central, and southern. ENERGY STAR windows must carry the NFRC label (discussed below), allowing comparisons of ENERGY STAR-qualified products on specific performance characteristics such as solar heat gain, insulating value, and infiltration.

■ Window Properties and NFRC Ratings

Before 1993, when NFRC developed standardized rating procedures for heat loss through windows, there was little consistency in how manufacturers listed energy performance. Today, energy-efficient windows are evaluated on five standardized performance measurements. These are labeled clearly on each window so you can compare one window to the next. NFRC labels describe the whole window U-factor (U-value), solar heat gain coefficient (SHGC), visible light transmittance, air leakage, and condensation resistance, all described briefly below. Because window dimensions and frame-to-glass ratio have a big impact on the total energy performance of a window, and yet it is difficult to account for all sizes, NFRC-rated windows fall into two representative sizes.

When a window manufacturer has its products certified, a great many different types and configurations of windows may be included. As defined by NFRC, a product line is a group of windows that have similar frame and operating characteristics (e.g., "Andersen Perma-Shield Casement windows"). Within each product line there may be various products. Differences among products include the number of layers of glazing, the use of or type of low-e coating, the type of gas-fill, edge spacer, etc.

For More Information:

For windows with tighter efficiency specifications than ENERGY STAR, the National Fenestration Rating Council (NFRC) keeps a directory of certified products on their website.
www.nfrc.org

U-value. U-value is the measure of the amount of heat (in Btus) that moves through a square foot of window in an hour for every degree Fahrenheit difference in temperature across the window. U-value is the inverse of R-value, which is familiar to many people as a measure of insulation thermal performance. Because it is the inverse, the lower the U-value rating, the better the overall insulating value of the window. Typical U-values range from 0.20 to 1.20. The U-factor ratings listed on NFRC labels (and in the NFRC Certified Products Directory) are whole-window U-values. That is, they take into account heat loss through the glass, window edge, and window frame.

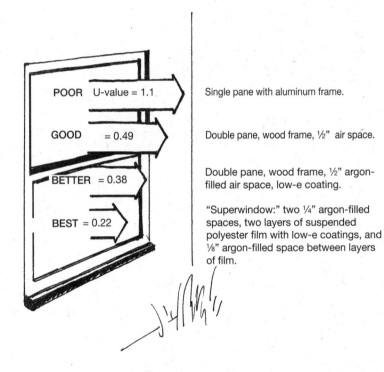

POOR U-value = 1.1 — Single pane with aluminum frame.

GOOD = 0.49 — Double pane, wood frame, ½" air space.

BETTER = 0.38 — Double pane, wood frame, ½" argon-filled air space, low-e coating.

BEST = 0.22 — "Superwindow:" two ¼" argon-filled spaces, two layers of suspended polyester film with low-e coatings, and ⅛" argon-filled space between layers of film.

The lower the U-value, the less heat escapes through a window. U-value quantifies thermal conductance in Btus per hour per square foot of window per °F (temperature difference between indoors and outside).

Solar heat gain coefficient (SHGC). The SHGC describes how much solar energy is transmitted through a window. Solar heat gain can be beneficial — providing free passive solar heat during the winter months — or it can be a problem, resulting in overheating during the summer. An SHGC of 0.8 means that 80% of the solar energy hitting the window gets through. As noted above, major window manufacturers make low-e windows with different solar heat gain coefficients. Windows with high coefficients (above 0.70) are designed for colder climates, while windows with low coefficients are designed for hotter climates.

Visible light transmittance. While SHGC describes the relative amount of solar energy that can pass through a window, the visible light transmittance is simply the relative amount of sunlight that can pass through, measured on a scale between 0 and 1. The higher the number, the greater the amount of light that can pass through. A tinted window or a window with a large ratio of frame to glass will have a lower visible transmittance value.

Air leakage. Another important energy property of windows is air leakage or air infiltration. Air leakage is already listed by many window manufacturers, in terms of cubic feet of air per minute per foot of crack. An optional air leakage value is included on NFRC labels and in the NFRC Certified Products Directory. The NFRC has adopted the same basic procedures for measuring air leakage that have been used by the industry in the past.

Condensation resistance. Finally, the ability of a window to resist the formation of condensation on the interior surface is very important in evaluating the relative durability of a window. The NFRC measures condensation resistance on a 0–100 scale. The higher the rating, the better that product is at resisting condensation formation. This rating is optional for new products, and it can not predict actual condensation.

National Fenestration
Rating Council®

CERTIFIED

World's Best Window Co.

Millennium 2000+

Vinyl-Clad Wood Frame
Double Glazing • Argon Fill • Low E
Product Type: **Vertical Slider**

ENERGY PERFORMANCE RATINGS

U-Factor (U.S./I-P)	Solar Heat Gain Coefficient
0.35	**0.32**

ADDITIONAL PERFORMANCE RATINGS

Visible Transmittance	Air Leakage (U.S./I-P)
0.51	**0.2**
Condensation Resistance **51**	**—**

Manufacturer stipulates that these ratings conform to applicable NFRC procedures for determining whole product performance. NFRC ratings are determined for a fixed set of environmental conditions and a specific product size. NFRC does not recommend any product and does not warrant the suitability of any product for any specific use. Consult manufacturer's literature for other product performance information.
www.nfrc.org

Look for the NFRC label to evaluate the performance of a high-efficiency window.

STATE OF THE ART

"Super-Insulation" Retrofit

Super-insulation retrofits refer to home improvements that go the extra mile to reduce the energy required to heat and cool a home. Such retrofits could involve the addition of 3–4 inches of rigid foam to exterior walls, boosting attic insulation to R-40 or more, replacement of windows with high-performance, low-e windows or the addition of low-e storm windows, thorough air tightening, and accounting for moisture problems. To achieve the aggressive 50% reduction in residential energy consumption that experts use as a target to prevent catastrophic climate change, this level of retrofit would be needed, especially in cold climates. Costs of such retrofits would be high (likely $50,000 per house) unless combined with other general remodeling.

Contact: BuildingGreen

www.buildinggreen.com

Green Roof Systems

Green roof systems are a way to insulate your roof that brings several added benefits. It may be something for you to consider if you have a low-slope or flat roof and are retrofitting your roof. Green roofs protect the roof membrane, reduce stormwater flows, and help green the built environment through rooftop plantings. They are good for energy efficiency and local temperature regulation because they both transmit less heat than conventional roofs and reflect or radiate less heat back into the local environment. Most roofs are not built to withstand the weight of a functional green roof, so installations on existing homes may require extra effort to shore up the roof.

Contact: Green Roofs for Healthy Cities

www.greenroofs.org

Chapter 3

Ventilation and Air Distribution

When you think about it, energy efficiency is not just about saving money on energy bills, it is really about using less energy to protect human health, assure comfort, and protect your house from damage. As air moves through your house, it removes pollutants that include odors, gases, particles, and (most surprisingly) moisture. But, it can also contribute to drafty walls and uncomfortable indoor temperature and humidity levels. Proper ventilation and air distribution play an important role in providing a safe, comfortable, and durable home as efficiently as possible.

What determines the air quality in your house? This chapter is designed to help answer this question by first explaining how air naturally moves through a typical house, then by describing what the main pollutants are and how best to control them.

For More Information:

To get the latest information on ventilation and air distribution and its impact on energy use, go to
www.aceee.org/consumerguide/ventilation.html.

How Air Moves Through Your House

■ Ventilation Basics

Virtually all houses, even yours, exchange indoor air with the outdoors. There are two reasons for this: First, a house will always have some leaks (air passages), however small, that connect the inside with the outside. These might include larger gaps around pipes, vents, and chimneys, and smaller cracks at places such as the join between the window frame and the wall. The second reason for air exchange is that there are temperature and pressure differences between inside and outside — air works hard to move from regions of high air pressure (warmer temperatures) to ones with lower air pressure (colder temperatures).

Consider a two-story house with a basement during winter. We all know that the warmest air tends to rise to the top floor, and that it can be several degrees warmer near the ceiling than in the basement. The warm, buoyant air is at a higher pressure than the cold air outside, and will want to move up and out, through windows, ventilation openings,

The tendency for air to move
upward through a house is
called the "stack effect."

and leaks in the walls, ceiling and roof. Downstairs and in the basement, low-pressure, cool air rushes in to replace the rising air in an attempt to maintain pressure balance. This natural upward current through the house is called the "stack effect." The same principle allows smoke and hot gases to rise up your chimney. In most houses, the amount of air that enters the house increases when it's windy and cold. When it's warm outside, the stack effect is much weaker or reverses.

Replacement of stale inside air with fresh, outside air is called ventilation. Ventilation can occur naturally, aided by the stack effect and open windows, or mechanically, with the use of a fan, or series of fans, that pull air in or out of the house. If air movement between inside and outside is accidental, we call it infiltration.

■ Air Distribution

Approximately two thirds of U.S. houses, including low-rise condos and townhouses, use forced air systems to move heating and cooling energy from a central furnace, air conditioner, or heat pump around the house using a duct system. This is not the same as ventilation — a forced air system is supposed to control how air is distributed within the house, not how air enters and exits. But as it turns out, your air distribution system is probably the largest source of infiltration. The furnace fan, or "air handler," will tend to move more than just the warm and cool air you want it to move. This is primarily because all conventional duct installations leak. A lot. Also, many houses keep the central equipment, along with leaky supply and return ducts, in uninsulated attic space where it is free to exchange all that carefully conditioned air with the outdoors.

What impact does all the leakiness have on indoor air quality and energy use? First, it makes it harder for the equipment to do its job. Excessive infiltration through the house and the ductwork causes the air conditioner to dehumidify more air on mild days, and run longer all the time. The furnace has to heat more air, too. The air filter faces an increased load of dirty outside air. Whether it's winter or summer, humidity control gets much harder because the winter air has too little water vapor load, and the summer air has too much. The only benefit is that all of the extra infiltration helps dilute local (indoor) pollutant sources that are otherwise not controlled well at their source.

Ideally, an energy-efficient and healthy house is able to carefully control the air that is coming in and going out, and to do so at just the right rate. Unfortunately, the age and construction method of your house, installation practices used, and the surrounding climate all complicate this mission. Newer houses are expected to be tighter, so they depend heavily on mechanical ventilation systems, which can be expensive to install or repair. Old houses allow more passive ventilation through air leakage, which usually keeps energy bills high. No matter what kind of ventilation strategies are possible in your home, perhaps the biggest variables in determining your indoor air quality and efficiency are what you choose to do in the space, and how you control pollution sources.

Understanding Indoor Air Pollutants

According to the Environmental Protection Agency, the top five air quality problems in the U.S. are all indoor air problems. Common residential indoor pollutants include excessive moisture, volatile organic compounds (VOCs), combustion products, radon, pesticides, dust particles, viruses, and bacteria. All of these are known to affect human health, and the resulting odors, dampness, stale air and stuffiness also make a house less comfortable. Most control strategies help with both gases and "particulates" like dust, pollen, and smoke particles. However, while filters of increasing sophistication can trap most nuisance and harmful particulates, there really aren't comparable removal strategies for gases. In some cases, activated charcoal and similar substances have been employed for this purpose.

■ Excess Moisture

Moisture is one of the most important and least recognized indoor pollutants, affecting both human health and the health of the building. The most common moisture problems arise when warm, moist air encounters a cool surface such as a mirror, window, or the wall of a cooler space. The air loses its capacity to carry moisture and the moisture condenses in droplets on the surface. Where moisture collects, so do mold, mildew, and dust mites, which can cause asthma or allergies, destroy wood products, and accelerate the rusting of metal building components. High indoor humidity can also facilitate "off-gassing" of toxins in furniture or cleaning products. Moisture affects comfort too: too much moisture in the cooling season creates a "cold-clammy" feeling inside. Too little moisture, which is common when it is very cold outside, can also be unhealthy and uncomfortable.

Water can be transported into a house as liquid, by driving rain or basement leaks; as vapor in humid air; or through capillary action—that is, the movement of moisture upward from the ground through porous materials like wood and concrete. Moisture also originates inside, from bath and kitchen activities, plants, and unvented gas appliances.

■ Radon

Radon is a radioactive gas that is generated naturally in the soil and enters the house from the ground. Radon is the second leading cause of lung cancer in the U.S. Its concentration in buildings varies regionally. There are excellent and relatively inexpensive ways to control radon concentrations where the levels are elevated. These generally involve using a small fan connected to a PVC pipe system to reduce the pressure of the soil gas and vent the radon-rich air harmlessly into the atmosphere so that it doesn't enter the house. The system pulls air from beneath the foundation slab, and generally exhausts harmlessly to the atmosphere at roof level.

For More Information:

Learn whether radon is a concern in your area or to test for elevated levels in your house.
www.nsc.org/ehc/radon/rad_faqs.htm

■ Combustion Products

Gas-fired appliances, including furnaces, water heaters, ranges and some dryers, produce carbon dioxide, carbon monoxide, nitrous oxides, and water vapor. If the appliance is not vented properly to the outside, or if there is enough negative air pressure around a non-sealed gas appliance to cause backdrafting, combustion pollutants can enter the house. Carbon monoxide (CO) is a particular concern because it is a colorless, odorless, toxic gas that is difficult to detect without a well-designed CO detector, and can be fatal. Likely sources of CO include unvented gas heaters, worn or poorly adjusted gas appliances and equipment, and improperly sized or damaged flues to furnaces or boilers. Auto, truck, or bus exhaust from attached garages, nearby roads, or parking areas can also be a source.

For More Information:

Learn more about the risk of CO at www.epa.gov/iaq/co.html

■ Volatile Organic Compounds (VOCs)

VOCs include a range of evaporated substances, including formaldehyde, which can be emitted by building materials and furnishings (such as furniture and carpets), gasoline from the garage, pesticides, and even cooking processes (the great scent of bread baking and the pungent smell of frying onions). VOCs include body odors, too. Many of these are just nuisances, but some, like formaldehyde, threaten health, sometimes at concentrations too low to be sensed.

■ Tobacco Smoke

Smoking is in a class by itself because its health consequences (even for non-smokers) are so well documented, and because it produces copious amounts of both harmful gases and particles. From both an indoor air quality and health perspective, few activities rival smoking in detrimental effects.

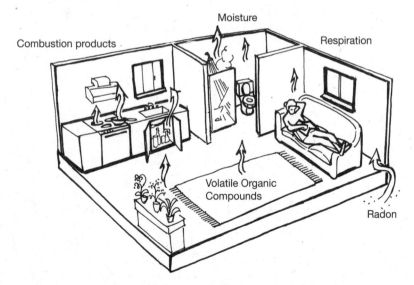

There are many potential sources of indoor air pollutants.

First Steps to Better Air Quality

■ Control the Sources First

Even before considering new ventilation equipment, you can control air quality to a surprising degree with fairly simple actions. Eliminating or reducing the source of the pollutant is almost always easier than capturing the pollutants after they have been released in the house. Banning indoor smoking is a clear example. Here are some other major ones.

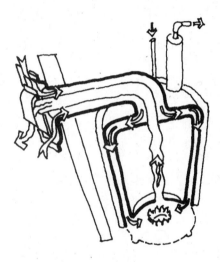

Use sealed combustion to control air intake and exhaust in a fuel-burning appliance.

Consider sealed combustion heating appliances. Combustion products from space or water heating appliances should never mix with the indoor environment. The terms sealed-combustion, direct-vent, and power-vented appliances all refer to appliances that vent their combustion products to the outside through a sealed pipe, under positive pressure. Sealed-combustion appliances are widely available and should always be used whenever located inside the conditioned space.

Check the garage connection. If your house has an attached garage, hire a professional to make sure the connection between your garage and your living space (including rooms above the garage) is airtight to prevent car exhaust and other chemicals you may keep in your garage from entering your house.

Control moisture. In many houses, construction and/or maintenance mistakes help water enter the house to make mischief. The most common and easily corrected errors come from improperly disposing of water shed from the roof. First, make sure that rain and gutters drain away from the building, not toward the basement or foundation (slab) walls.

Grade the ground so it slopes away from the house. Water that collects near the foundation is very likely to leak in, at least during heavy rains. In southern climates in particular, dehumidifiers may be necessary to control moisture levels in the house, if the air outside is hot and humid.

If condensation is a problem, pay attention to how you ventilate areas that exchange a lot of air with the outside or produce a lot of moisture. Clearly activities that produce a lot of steam, like showers, should be vented directly to the outside. Other scenarios are less obvious. For instance, ventilating a basement in the summer will pull hot, humid air inside, causing condensation on the cold interior wall. Similarly, if your attic is vented, condensation may develop on the underside of your roof when the temperature drops at night, particularly if there is a source of warm moist air in the attic, either from excessive air leakage in the ceiling or heating equipment located in the attic. During the winter, this could lead to massive icicles forming along the roof edge. If it's cold outside and there's a break in insulation, warm, moist air from inside will condense along that cold seam, especially in rooms that contain plants or appliances that generate water vapor. Read up on how to make sure these areas of your house are well sealed and insulated in Chapter 2.

Use safe household products. Use pesticides and cleaning agents wisely and store them safely. Choose paints that are water-based instead of oil-based to reduce VOCs. Choose building materials and finishes that are known to reduce pollutant emissions.

■ Exhaust Locally, at the Source
If you can't control the pollutant source, then exhaust it from the building where it is produced. It takes a lot less energy to remove a small amount of moisture-laden bathroom or kitchen air than a larger volume

For More Information:

Information about household products that pose fewer health risks can be found here.

www.thegreenguide.com   ■   www.watoxics.org   ■   www.usgbc.org

of air from the whole house. When showering or cooking use an exhaust fan to exhaust pollutants directly to the outdoors before they can negatively impact air quality in other rooms. Typically, this can be done with individual exhaust fans in bathrooms and a vent hood over the kitchen stove. Opening a window may not be enough to clear the air of moisture and other pollutants, especially on warm, calm, or humid days.

If you are looking to install a new exhaust fan, look for an ENERGY STAR-rated fan. ENERGY STAR fans are energy-efficient, provide the right airflow, and are also quiet — about as loud as a modern refrigerator. To provide the right ventilation without wasting energy, fans should be 50 cfm for bathrooms 100 ft^2 or less. Kitchen range hoods are available in many sizes, but typically an up-draft range hood should provide 100 cfm of ventilation for wall-mounted hoods and 150 cfm for island hoods. They should always have a ducted passage to the outside that is fitted with a damper to prevent infiltration.

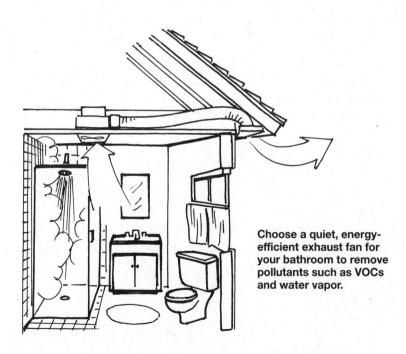

Choose a quiet, energy-efficient exhaust fan for your bathroom to remove pollutants such as VOCs and water vapor.

If you have a bathroom fan but you are concerned about forgetting to use it or leaving it on too long, you may want to consider hiring an electrician to install an occupancy sensor, humidity sensor, or other controls that turn the fan on with the light switch.

■ Filter Out Particulates

In addition to removing various gases by the methods discussed above, you may also have to deal with particulates such as soot, dust, pollen and tobacco smoke. These pollutants cannot be handled with exhaust alone, just as filters don't do the best job of removing gases. Mechanical systems that circulate and filter air through the house are the best approach.

Mechanical filters. There are two basic kinds of air filters. Mechanical filters trap particles on the surfaces of porous "media" that act like three-dimensional sieves. The designs vary in shape and the size of the smallest particles they trap well, as measured by "MERV," or "Minimum Efficiency Report Value." MERV 4 or 5 is pretty characteristic of residential air filters. MERV 8 or 9 will do a good job on particles down to about 3 microns in size. A fraction of fine smoke and bacteria will even pass through these. MERV 12 is probably higher than needed for any residential use other than for very specialized clinical needs. In general, mechanical filters that remove finer particles cost more to purchase and have substantially higher resistance to airflow; they require more fan electricity. This may amount to hundreds of kilowatt-hours per year, or over $100 in some regions.

Electrostatic filters. "Electronic," "electrostatic," and "electret" air filters use a different physical principle: particles passing through these assemblies are attracted to surfaces that have an electric charge. They are trapped there until the filter is cleaned or discarded. In contrast to mechanical filters, these usually pose little resistance to airflow. However, "permanent" types do require regular cleaning to maintain their performance.

All forced air systems need filters, and all filters need regular cleaning or replacement. The first thing is to be sure that you know where your filter is. Usually, it is in a slot, often protected by a door, near the connection between the return air and the air handler for the furnace or heat pump. Many attic-based systems will have their filters installed in

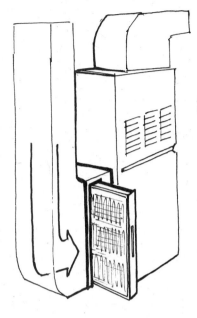

Replace furnace air filter monthly during the heating season.

return air registers, instead, for easier access. This may mean multiple filters, one for each return. Most filters are disposable, with expected lives of a month, several months, or a year. Dirty and clogged air filters require much more fan energy, and may lead to inadequate air delivery, or shorten the life of expensive components.

■ General Dilution

After local exhaust is applied to remove concentrated pollutants at their source and any particulates are filtered from the air, it is recommended that houses always have some planned ventilation strategy that dilutes the general body of lingering pollutants by mixing stale air with new air and moving it through the house. This can be continuous or periodic, and it may be achieved simply by opening windows. Throughout the day, or if there isn't much breeze, operating a 100 cfm fan would continuously provide adequate ventilation for a 1,500 square foot house with 8-foot ceilings. The same result could be obtained with several smaller fans in different rooms, or with a larger fan that ran intermittently.

Signs That You May Have an Air Quality Problem

Just like our bodies, houses will present symptoms that help us diagnose when the system isn't working properly. The main problems to diagnose are excess infiltration (through walls or ducts), inadequate air exchange (too little ventilation), and excess moisture. Often, excess moisture speaks for itself. If you see condensation, mold, or excessive ice build-up, you know you have a problem. (See the section "control moisture.")

■ Signs of Excess Infiltration

Excess air infiltration often leaves characteristic "fingerprints" that can be read or felt. Most obvious is when you feel a draft, particularly in winter and especially when you are closer to exterior walls. One less obvious sign of infiltration is frequent static electricity discharges in winter due to indoor humidity levels that are too low. Also look for "ghost" tracks of dark dust near boundaries that are poorly sealed. These tracks are most common on light carpets near exterior walls, and as black edges on fiberglass insulation batts between joists in the attic. What is happening is that dusty air is moving through the wall and dropping the dust when it reaches the wall and its velocity slows.

If you have air conditioning and heating equipment in the attic, your ductwork is very likely the cause of infiltration, and you should seal and insulate the ducts accordingly. Otherwise, the best way to address excess infiltration is to seal obvious leaks in your windows, walls, and connections between your house and your attic, as described in Chapter 2. If you are unsure of how to find leaks, hire a contractor to conduct a blower-door test to identify and seal them.

■ Signs of Inadequate Air Exchange

Odors can be very good indicators of too little ventilation and may also reveal how air travels through your house. How persistent are certain odors? Where are they generated and where are they getting trapped? If odors persist for hours, it is because either the source keeps on releasing the chemicals we smell, or the air carrying the odors does not have a quick, controlled way out of the house (if your house contains a lot of cloth furnishings, they will also trap odors). It can be instructive to think through what it means when some strong odor does or does not percolate through the house. Consider the onions frying in the kitchen. If that odor doesn't reach the rest of the house, this may mean that kitchen ventilation is working very well. In this case, there should be relatively little onion fragrance in the kitchen, too. But what if the kitchen proclaims, "Onions cooking," but no one notices in the far bedrooms? That would suggest that air is not circulating throughout the house very effectively.

■ Signs of Too Much Air Exchange

If you live in a dry climate (or a place with cold, dry winters), and it is uncomfortably dry inside, check the fan speed on any ventilation equipment you might have installed before running out to buy a humidifier. Likewise, if you live in a hot, humid climate, high ventilation rates can increase indoor humidity beyond what the air conditioner can control on its own, requiring supplemental dehumidification.

Whole-House Ventilation Strategies for New Houses

Conventional wisdom among builders of modern, high-performance homes in most North American climates is to build as tight as possible, and then ventilate with a well-designed mechanical system. Even though this uses a little energy, you are likely to save more heating and cooling energy from having a tight house than you will use to ventilate. Mechanical ventilation is achieved in one of three ways: exhaust-only, supply-only, and balanced. To varying degrees, all of these strategies are more economically installed in new houses and more difficult to do as retrofits, because access in existing houses may be poor.

■ Exhaust-Only

Exhaust fans (kitchen, bathroom, and/or whole house fans) tend to "depressurize" the building, causing infiltration of outside air through any cracks or openings it can find. In the North, where winter is more intense than summer, exhaust-only ventilation may be adequate without inviting damage from moisture. Because summers tend to be short and moderate in cold climates, except for a few days, the building is unlikely to be damaged by occasionally drawing in very hot and humid air through the structure. Conversely, exhaust-only ventilation strategies should not be used in the South. If hot and humid air is drawn into the building for months on end, condensation, mold, and damage are likely to develop.

■ Supply-Only

Supply ventilation systems draw clean outside air into the interior living space, usually through a supply vent that feeds into the return duct of a forced air system. Advantages of supply-only ventilation include the ability to control where incoming air is coming from, treat the incoming air, and minimize humid air that is pulled into the living space. Controlled supply also minimizes the potential for combustion appliance backdrafting, a dangerous type of uncontrolled infiltration that is more common in well-sealed and poorly vented basements. Supply-only strategies will "pressurize" the house, which keeps moisture out in hot, humid climates but may induce drafts in cold climates as warm air escapes to the outside.

■ Balanced

Balanced whole-house ventilation systems exhaust indoor air and supply outdoor air in roughly equal amounts. This way the pressure of the interior space stays relatively constant, although this is rarely perfect. Balanced ventilation is essentially a well-controlled combination of the exhaust and supply strategies discussed above, but it takes a very tight house and good engineering. Often, a balanced system involves a powered heat recovery or energy recovery ventilator (HRV or ERV) that improves efficiency and pressure balance by exchanging energy (from temperature and humidity differences) between the outgoing and incoming airstreams. HRVs transfer only sensible heat while ERVs also transfer heat from humid air. The relatively large electric power use of ERVs and HRVs generally make them a luxury option in mild climates and should not be considered as an economic option for most existing buildings.

Whole-House Ventilation Strategies for Old Buildings

For old buildings, creating the kind of tight construction that would make installing a mechanical ventilation system worthwhile is probably difficult and economically risky. If you live in an old building, the first question to ask is whether a whole-house ventilation system is needed at all. Is your house displaying any symptoms? What are your main comfort concerns? Have you conducted a radon test? If you have limited complaints, it would be most cost-effective to simply take the "first steps:" control sources, exhaust local sources, and keep your filters clean and your ducts tight.

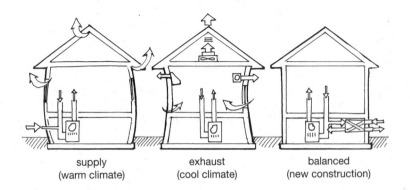

supply
(warm climate)

exhaust
(cool climate)

balanced
(new construction)

Mechanical ventilation involves some combination of exhaust and supply fan power. A "balanced" system implies more control and integration of supply and exhaust.

■ Local Exhaust

If you live in a hot-humid climate, exhaust-only ventilation can depressurize the house. This could drive humid outdoor air into the wall cavities, where it can condense and cause grave mischief. It is better to consciously introduce outside air into the house than have it infiltrate through the walls. However, you're unlikely to get into trouble by using exhaust fans in moderation, and by opening the windows occasionally when it feels stuffy or the exhaust fans have been on for a long time.

■ Managing your Current System

If you have an existing forced-air system, it is likely that your thermostat has a switch for controlling the furnace fan. In "auto" mode, the fan runs when the furnace or air conditioner (or heat pump) is running, and for a short time afterwards. When the switch is set to "on," the fan runs continuously. Full-time fan operation will certainly move more air through the filter, and encourage more even distribution of conditioned air throughout the building, but it is not a good idea because it will waste a large amount of energy. In the air conditioning mode, continuous fan operation is a really bad idea, and should be avoided. When the air conditioner is operating, it removes humidity by cooling the air to perhaps 50°F to 55°F, cold enough that the ability of the air to hold moisture is very low. So, the water condenses out, onto the coils and fins of the air conditioner evaporator. If the fan stops soon after the condensing unit shuts off the compressor, most of this water drips off the evaporator and goes out through a drain line. However, if the fan runs continuously, that water is instead re-evaporated and circulated through the house. This leads to cold-clammy houses.

■ Natural Ventilation

Natural ventilation is usually employed as a cooling strategy, but in principle, the idea is to replace stuffy indoor air with cool outdoor air. In order for it to be most effective, the incoming air should be cooler and dryer than the inside air, making this strategy most effective in milder climates, at night, or on cooler, drier days. Keep the house closed up on hot days and try to limit unwanted heat gains and then ventilate the house at night. In breezy locations, natural ventilation can be provided simply by opening screened windows. Plantings and fences can be used to help funnel breezes towards your house. If there isn't much wind, you'll need to provide mechanical ventilation with either window fans or a whole-house fan. These are explained in Chapter 5 on cooling.

■ Whole-House or Room Dehumidifiers and Humidifiers

There are many situations, particularly in hot climates, in which conventional air conditioners do not remove enough moisture. This is particularly true on very humid days with moderate temperatures (in the 70s). Under these conditions, the air conditioner probably will not run enough to remove the moisture load. New dehumidifiers are now available that install in place of a section of central ductwork, so they dehumidify (but reheat a bit) all the air circulating in the house. These are more expensive than room dehumidifiers, but may be very effective when nothing else works.

Excessive and continuing static electricity build-up in winter is typically taken as a sign that the relative humidity in the house is too low and should be increased. With forced air systems, this is generally accomplished by adding a humidifier that injects water vapor — or very fine water droplets — into the supply air near the furnace. Before taking this step, remember that a very dry house through the winter is generally a sign of excessive infiltration. See if you can reduce the leakage, thereby saving energy and increasing comfort.

Chapter 4

Heating Systems

Heating is the largest energy expense in most homes, accounting for 35% to 50% of annual energy bills. Reducing your energy use for heating provides the single most effective way to reduce your home's contribution to global environmental problems. Heating systems in the United States spew over a billion tons of CO_2 into the atmosphere each year and about 12% of the nation's sulfur dioxide and nitrogen oxides.

A combination of conservation efforts and a new high-efficiency heating system can often cut your pollution output and fuel bills by one-third, and in some homes by half. Just upgrading your furnace or boiler to a high efficiency model can save 1–2 tons of CO_2 emissions each year in colder climates.

For More Information:

To get the latest information on reducing the energy you use for heating, go to aceee.org/consumerguide/heating.html.

Heating System Basics

A heating system replaces heat that is lost through the shell of your house. How much energy your heating system requires to replace that lost heat depends on four factors: where the house is located (in colder places, the house will lose more heat); how big the house is; the energy efficiency of the house; and how energy-efficient the heating system is.

You can't do much about the first factor. All other things being equal, the bigger the house, the more energy it will take to heat it. Chances are, you're not going to decrease the size of your house.

As for the efficiency of your house, improving insulation, sealing air leaks, and repairing the heat distribution system (ducts or pipes) represent excellent opportunities for saving energy and dollars. These measures are addressed in Chapter 2.

Did You Know?

Heat is measured in British thermal units, or Btu.
- kitchen match = 1 Btu
- 1 kilowatt-hour = 3,413 Btu
- 1 therm of gas = 100,000 Btu

A typical house will use tens of thousands of Btus per hour on a cold day.

Whether or not you've buttoned up your house, you can probably save a great deal by upgrading your heating system, either by installing a new high-efficiency system or by boosting the efficiency of your present system. Both of these options are addressed in this chapter.

But first, when considering the various options for improving or replacing your heating system, it helps to know some of the lingo. A lot of confusing terms and concepts are thrown around by salespeople or heating system technicians, and you don't want to get left behind. Central heating systems have three basic parts: the heating plant itself where fuel is converted into useful heat, a distribution system to deliver heat to where it is needed, and controls to regulate when and how the system runs and when it turns off.

Central Heat

■ Furnaces

The majority of North American households depend on a central furnace to provide heat. A furnace works by blowing heated air through ducts which deliver the warm air to rooms throughout the house via air registers or grills. This type of heating system is called a ducted warm-air or forced warm-air distribution system. It can be powered by electricity, natural gas, or fuel oil.

Inside a gas- or oil-fired furnace, the fuel is mixed with air and burned. The flames heat a metal heat exchanger where the heat is transferred to air. Air is pushed through the heat exchanger by the "air handler's" furnace fan and then forced through the ductwork downstream of the heat exchanger. At the furnace, combustion products are vented out of the building through a flue pipe. A typical furnace has a hard time keeping valuable heat from escaping up the flue. "Condensing"

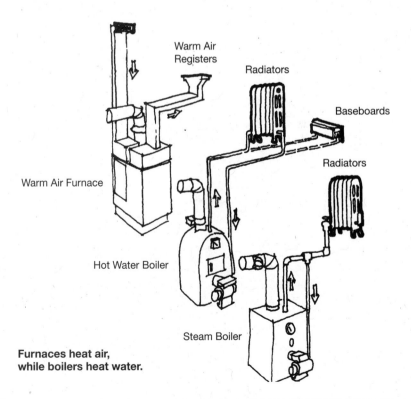

Warm Air Registers

Radiators

Baseboards

Radiators

Warm Air Furnace

Hot Water Boiler

Steam Boiler

**Furnaces heat air,
while boilers heat water.**

furnaces are designed to reclaim much of this escaping heat by cooling exhaust gases to below 212°F, where water vapor in the exhaust condenses into water. This is the primary feature of a high-efficiency furnace (or boiler).

Heating system controls regulate when the various components of the heating system turn on and off. The most important control from your standpoint is the thermostat, which turns the system — or at least the distribution system — on and off to keep you comfortable. A typical forced air system will have a single thermostat. But, there are other internal controls in a heating system, such as "high limit" switches that are part of an invisible but critical set of safety controls.

The efficiency of a fossil-fuel furnace or boiler is a measure of the amount of useful heat produced per unit of input energy (fuel). Combustion efficiency is the simplest measure; it is just the system's

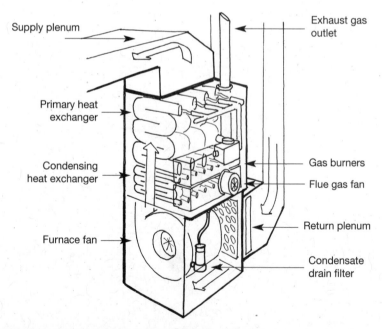

Supply plenum

Exhaust gas outlet

Primary heat exchanger

Condensing heat exchanger

Gas burners

Flue gas fan

Return plenum

Furnace fan

Condensate drain filter

Condensing furnaces have a second heat exchanger that recaptures heat from excess water vapor.

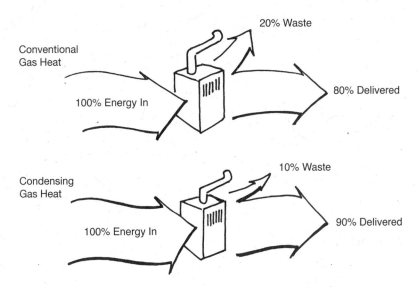

The best gas furnaces and boilers today have efficiencies over 90%.

efficiency while it is running. Combustion efficiency is like the miles per gallon your car gets cruising along at 55 miles per hour on the highway.

Because furnaces and boilers lose heat when they cycle "off," an estimate of seasonal efficiency is more commonly reported in manufacturer literature and used to set minimum efficiency standards. This is called annual fuel utilization efficiency, or AFUE. AFUE accounts for start-up, cool-down and other operating losses that occur in real operating conditions. AFUE is like your car mileage between fill-ups, including both highway driving and stop-and-go traffic. The higher the AFUE, the more efficient the furnace or boiler.

■ Boilers

Boilers are special-purpose water heaters. While furnaces carry heat in warm air, boiler systems distribute the heat in hot water, which gives up heat as it passes through radiators or other devices in rooms throughout the house. The cooler water then returns to the boiler to be reheated. In a hot-water system, also called a "hydronic" system, the water is typically heated to about 180°F (or less in a well-designed system).

In steam boilers, which are much less common in homes today, the water is boiled and steam carries heat through the house, condensing to water in the radiators as it cools. Oil and natural gas are commonly used; as with gas- and oil-fired furnaces, boilers can be designed to condense water vapor in the exhaust pipe to reclaim escaping heat.

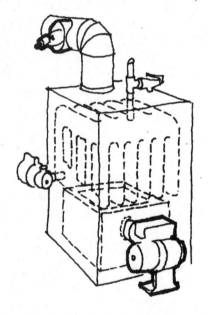

Boilers work like furnaces, except that they heat water instead of air.

Instead of a fan and duct system, a boiler uses a pump to circulate hot water through pipes to radiators. Some hot-water systems circulate water through plastic tubing in the floor, a system called radiant floor heating (see "State of the Art Heating"). Important boiler controls include thermostats, aquastats, and valves that regulate circulation and water temperature. Although the cost is not trivial, it is generally much easier to install "zone" thermostats and control for individual rooms with a hydronic system than with forced air. Some controls are standard features in new boilers while others can be added on to save energy (see Modifications by Heating System Technicians section, p. 89).

■ Heat Pumps

Heat pumps are just two-way air conditioners (see detailed description in Chapter 5). During summer, an air conditioner works by moving heat from the relatively cool indoors to the relatively warm outside. In winter,

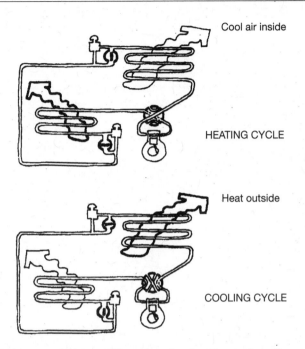

Cool air inside

HEATING CYCLE

Heat outside

COOLING CYCLE

Heat pumps can work in two different modes: heating and cooling.

the heat pump reverses this trick, scavenging heat from the cold outdoors with the help of an electrical system, and discharging that heat inside the house. Almost all heat pumps use forced warm-air delivery systems to move heated air throughout the house.

There are two relatively common types of heat pumps. Air-source heat pumps use the outside air as the heat source in winter and heat sink in summer. Ground-source (also called geothermal, GeoExchange, or GX) heat pumps get their heat from underground, where temperatures are more constant year-round. Air-source heat pumps are far more common than ground-source heat pumps because they are cheaper and easier to install. Ground-source heat pumps, however, are much more efficient, and are frequently chosen by consumers who plan to remain in the same house for a long time, or have a strong desire to live more sustainably. How to determine whether a heat pump makes sense in your climate is discussed further under "Fuel Options".

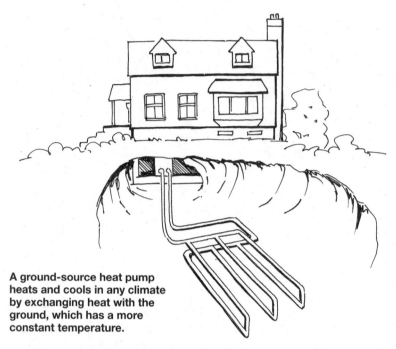

A ground-source heat pump heats and cools in any climate by exchanging heat with the ground, which has a more constant temperature.

Whereas an air-source heat pump is installed much like a central air conditioner, ground-source heat pumps require that a "loop" be buried in the ground, usually in long, shallow (3–6' deep) trenches or in one or more vertical boreholes. The particular method used will depend on the experience of the installer, the size of your lot, the subsoil, and the landscape. Alternatively, some systems draw in groundwater and pass it through the heat exchanger instead of using a refrigerant. The groundwater is then returned to the aquifer.

Because electricity in a heat pump is used to move heat rather than to generate it, the heat pump can deliver more energy than it consumes. The ratio of delivered heating energy to consumed energy is called the coefficient of performance, or COP, with typical values ranging from 1.5 to 3.5. This is a "steady-state" measure, and not directly comparable to the heating season performance factor (HSPF), a seasonal measure mandated for rating the heating efficiency of air-source heat pumps. Converting between the measures is not straightforward, but ground-source units are generally more efficient than air-source heat pumps. (Cooling is addressed in Chapter 5.)

Direct Heat

■ Gas-Fired Space Heaters

In some areas, gas-fired direct heating equipment is popular. This includes wall-mounted, free-standing, and floor furnaces, all characterized by their lack of ductwork and relatively small heat output. Because they lack ducts, they are most useful for warming a single room. If heating several rooms is required, either the doors between rooms must be left open, or another heating method is necessary. Better models use "sealed combustion air" systems, with pipes installed through the wall to both provide combustion air and carry off the combustion products. These units can provide acceptable performance, particularly for cabins and other buildings where large temperature differences between bedrooms and main rooms are acceptable. The models can be fired with natural gas or propane, and some burn kerosene.

■ Unvented Gas-Fired Heaters: A Bad Idea

Gas or kerosene space heaters that *do not* have an exhaust vent have been sold for decades, but we strongly discourage their use for health and safety reasons. Known as "vent-free" gas heating appliances by manufacturers, they include wall-mounted and free-standing heaters as well as open-flame gas fireplaces with ceramic logs that are not actually connected to a chimney.

Manufacturers claim that because the products' combustion efficiency is very high, they are safe for building occupants. However, this claim is only valid if you keep a nearby window open for adequate fresh air—which defeats the purpose of supplemental heat. Dangers include exposure to combustion byproducts, as discussed in Chapter 3, and oxygen depletion (these heaters must be equipped with oxygen depletion sensors). Because of these hazards, at least five states (California, Minnesota, Massachusetts, Montana, and Alaska) prohibit their use in homes, and many cities in the United States and Canada have banned them as well.

■ Electric Space Heaters

Portable (plug-in) electric heaters are inexpensive to buy, but costly to use. These resistive heaters include "oil-filled" and "quartz-infrared" heaters. They convert electric current from the wall socket directly into heat, like a toaster or clothes iron. As explained further under "Selecting A New System," it takes a lot of electricity to deliver the same amount of useful heat that natural gas or oil can provide on site. A 1,500 watt plug-in heater will use almost the entire capacity of a 15-amp branch circuit; thus, adding much additional load will trip the circuit breaker or blow the fuse. The cost to operate a 1,500 watt unit for an hour is simple to compute: it is 1.5 times your electricity cost in cents per kilowatt-hour. At national average rates—9.5¢ kWh for electricity—that heater would cost 14¢ per hour to run—and quickly cost more than its purchase price. On the other hand, for intermittent use, it is the "least-bad" solution when alternatives would require major investments to improve ductwork for a specific area, for example. Just remember, electric resistance heat is usually the most expensive form of heat and it is, therefore, seldom recommended.

"Electric baseboard heat" is yet another kind of resistive heating, similar to a plug-in space heater except that it is hard-wired. It has two principal virtues: the installation cost is low, and it is easy to install individual room thermostats so you can turn down the heat in rooms that aren't being used. Operating costs, as for all resistive systems, are generally very high, unless the house is "super-insulated."

■ Wood-Burning and Pellet Stoves

Wood heating can make a great deal of sense in rural areas if you enjoy stacking wood and stoking the stove or furnace. Wood prices are generally lower than gas, oil, or electricity. If you cut your own wood, the savings can be large. Pollutants from wood burning have been a problem in some parts of the country, causing the U.S. Environmental Protection Agency (EPA) to implement regulations that govern pollution emissions from wood stoves. As a result, new models are quite clean-burning. Pellet stoves offer a number of advantages over wood stoves. They are less polluting than wood stoves and offer users greater convenience, temperature control, and indoor air quality. This edition of the *Consumer Guide to Home Energy Savings* does not include detailed information on wood and pellet stoves, because independent and reliable comparative rating information is not generally available.

■ Fireplaces

Gas (and most wood) fireplaces are basically part of a room's décor, providing a warm glow (and a way to dispose of secret documents), but

Should I Replace My Heating System?

Less than 10 years old **NO**	10–20 years old **DEPENDS**	More than 20 years old **YES**
↓	↓	↓
See "Upgrading your Heating System"	*Follow Chart Below* ↓	*See "Selecting a New Heating System"*

Current System: GAS or OIL FURNACE or BOILER

System Condition	Climate	Replace?
Pre-1992 standing pilot	Any region	**YES.** *See "Selecting a New Heating System".*
1992 or later AND less than 88 AFUE	Cold or moderate climate	Consider replacement rather than any major repair. *See "Selecting a New Heating System".*
1992 or later AND less than 88 AFUE	Very mild climate (e.g., deep South, Pacific NW)	**NO.** Research options so you are prepared if system fails or requires major repairs.

Current System: ELECTRIC HEAT PUMP

System Condition	Climate	Replace?
Pre-1992	Any region	**YES.** Equipment near end of service life. Replacement will save energy for heating and cooling.
1992 or later AND less than 7.0	Any region	Consider replacement if major component breaks. *See "Selecting a New Heating System".*

Current System: ELECTRIC RESISTANCE

System Condition	Climate	Replace?
Furnace Any AFUE	Any region except the deep South and PNW	**YES.** *See "Selecting a New Heating System".*
Baseboard	Any region except the deep South and PNW	**YES.** *See "Selecting a New Heating System".*

typically not an effective heat source. With customary installations that rely on air drawn from the room into the fireplace for combustion and dilution, the fireplace will generally lose more heat than it provides, because so much warm air is drawn through the unit and must be replaced by cold outside air. On the other hand, if the fireplace is provided with a tight-sealing glass door, a source of outside air, and a good chimney damper, it can provide useful heat.

Should I Replace My Existing Heating System?

This can be a difficult question. If your furnace or boiler is older than 20 years, chances are it is probably oversized, inefficient, and likely to fail soon (a typical forced air furnace will last less than 25 years, though some boilers can last twice that long). With the guidance of a good contractor, replacing it with a modern high-efficiency model is a good investment. Old coal burners that were switched over to oil or gas are prime candidates for replacement, as are gas furnaces without electronic (pilotless) ignition or a way to limit the flow of heated air up the chimney when the heating system is off (vent dampers or induced draft fan).

If your furnace or boiler is 10–20 years old, the decision to replace it depends on its age, size, condition and performance. If you can't seem to stay comfortable or keep your heating bills down, the best first step would be to hire a highly-qualified home performance or heating contractor who can help you evaluate your existing system within the context of your home and determine the best actions to take. In some cases they may find that your furnace isn't the problem, but rather your ceiling insulation, windows, or ducts need some help. It may be that a component of your system was improperly installed or needs a good tune-up by a certified technician (more about that later in this chapter). Chapter 2 provides more information on hiring home performance contractors and energy auditors who can help.

■ Are the Savings Worth It?

If you heat with electric resistance heat, rising electricity prices may force you to switch to a gas, oil, or heat pump system that is more affordable. But it is important to carefully evaluate these fuel tradeoffs (read more under "Fuel Options"), because fuel switching can be expensive.

If you currently have a gas- or oil-fired furnace or boiler and know its rated efficiency, it's pretty easy to calculate the savings you will get by replacing it with a more efficient system of the same type, assuming both the old and new systems are correctly installed. The chart below will help you determine the potential savings resulting from replacement of your existing system. If you don't know the AFUE of your system, ask your heating service technician or energy auditor.

The numbers in the chart assume that both the old and new systems are sized properly; savings will be greater than indicated if the old system is too large. Also, keep in mind that other system deficiencies will impact your overall savings. For example, most duct systems are not properly installed; unless you've had your ducts checked, 20% of the heat generated by your furnace is likely wasted through duct leakage. If so, it would be more cost-effective to put your money where it counts and make sure your ducts are repaired and sealed by a highly-trained contractor. Depending on how old your system is, you may be able to use the rest of the money you would have spent to make a more modest efficiency upgrade to get the same level of savings.

TABLE 4.1 Dollar Savings per $100 of Annual Fuel Cost

		AFUE of new system			
		80%	85%	90%	95%
AFUE of existing system	50%	$38	$41	$44	$47
	55%	31	35	39	42
	60%	25	29	33	37
	65%	19	24	28	32
	70%	13	18	22	26
	75%	6	12	17	21
	80%		6	11	16
	85%			6	11

To estimate savings using the table, find the horizontal row corresponding to the old system's AFUE, then choose the number from that row that is in the vertical column corresponding to the new system's AFUE. That number is the projected dollar savings per hundred dollars of existing fuel bills. For example, if your present AFUE is 65% and you plan to install a high-efficiency natural gas system with an AFUE of 90%, then the projected saving is $28 per $100. If, say, your annual fuel bill is $1,300, then the total yearly savings should be about $28 x 13 = $364. If you expect fuel prices to continue rising, you would expect even larger savings in the future, and vice versa.

That's a lot of money to save each year, especially when you consider the expected lifetime of a heating system, but it still doesn't answer the question of whether or not replacing the system is a good investment. To answer that, you can calculate the first year return on investment (ROI). The equation is as follows:

ROI = first year savings ÷ installed cost

ROI = $364 ÷ $2,500 = 0.14 = 14% *— tax free!*

A 14% return is pretty good — much higher than what you receive from a savings account or certificate of deposit. Plus, unlike most other investments, energy conservation investments are tax free. If fuel prices go up, the annual savings and return on investment also go up. For example, if fuel prices increase 30%, the annual savings in this example increases to $473, and the return on investment increases to 19%.

While you're doing the math, don't forget that some states or utilities (or even the federal government) may offer tax incentives or rebates that will reduce the purchase price of high-efficiency heating equipment. Be sure to check with your local utility and state energy office to see what incentives are available.

Selecting a New Heating System

So you've decided to buy a new heating system. While it is useful to understand what fuel, distribution, and equipment options you want to consider, the first step in selecting a new system is finding a knowledgeable and reliable contractor who can help you navigate the choices outlined below. It is paramount that the system is properly installed to ensure safety, reliability, and maximum efficiency.

Selecting a Heating System

To ensure a reliable, high-efficiency system, you must first find a skilled contractor with experience in high-efficiency heating systems. Walk through these options with your contractor as they may apply to your particular home. (See "Choosing a Heating System Contractor.")

Current System	Replacement Options	Recommendations
Current System: GAS		
Forced Air	Condensing Furnace	AFUE ≥ 90 (ENERGY STAR) High-efficiency furnace fan (p. 86) Sealed-combustion preferable May require new flue lining for water heater or new power vent water heater (p. 129)
	Non-Condensing Furnace	Mild climates (deep South, Pacific NW) only. High-efficiency furnace fan (p. 86)
	Switch to Hydronic System	Expense may preclude unless part of large-scale renovation Better option where mild summers make central air conditioning unnecessary
Hydronic	Non-Condensing Boiler	Mild climates (deep South, Pacific NW) only
	Condensing Boiler	AFUE ≥ 85 (ENERGY STAR) Sealed-combustion preferable Insist on outdoor reset or equivalent controls (p. 90) Consider indirect water heater tank (p. 134)
	Switch to Forced Air	Consider if you have single pipe steam system and central air conditioning
Combination Space/Water Heater		Explore this if you also are looking to replace your water heater (p. 133)

Selecting a Heating System *continued*

To ensure a reliable, high–efficiency system, you must first find a skilled contractor with experience in high–efficiency heating systems. Walk through these options with your contractor as they may apply to your particular home. (See "Choosing a Heating System Contractor.")

Current System	Replacement Options	Recommendations
Current System: OIL		
Forced Air	Furnace	AFUE ≥ 83 (ENERGY STAR)
Hydronic	Boiler	AFUE ≥ 85 (ENERGY STAR) Sealed-combustion preferable Insist on outdoor reset or equivalent controls (p. 90) Consider indirect water heater tank (p. 134)
Switch to Gas	(see Gas options)	Gas may save you money and allow for a more efficient system
Current System: ELECTRIC		
Resistance: Furnace or Baseboard System	Switch to Heat Pump	Cheaper option especially if you have forced air and are replacing a central air conditioner
	Switch to Gas	Consider if you have a gas line and a central air conditioner in good condition
	Supplemental Direct Heat	Not recommended as replacement option. Only use in rooms that are remote from the central system
Heat Pump	High-efficiency Air-Source Heat Pump	ENERGY STAR or better, quality installation is important
	Ground-Source Heat Pump	ENERGY STAR or better, quality installation is important Specify integrated water heating (p. 133)

■ Choosing a Heating System Contractor

You'll spend a lot of money to have a new furnace or boiler installed — anywhere from about $2,000 for a simple furnace installation to over $5,000 for a complicated boiler installation. A complete new system, with a new air conditioner and routine duct improvements, may cost over $7,000 for a proper job with good equipment, at least in high labor-cost regions. Any modifications to the distribution system will add significantly to the cost. To get the best deal, you should get bids for the price of equipment and installation from several contractors. It isn't unusual for bids to differ by as much as $1,000, so get at least two or three bids.

Try not to let the lowest price be the main reason for selecting a contractor. Better contractors might charge more, but they may offer greater value. When evaluating bids also look at what you get for the price: quality, energy savings, and warranty. Extremely low bids may not include all routine services and customary warranties. Ask the contractor if he or she has had any special training in high-efficiency equipment. A well-trained, up-to-date contractor will not try to discourage you from purchasing high-efficiency equipment. If you think your old heating system is covered with asbestos insulation, make sure the contractor knows about it and will follow the proper procedures to deal with the asbestos. If you are not familiar with the contractor, ask for customer references and follow up on them. Make sure the contractor is fully bonded and insured.

A good bid should be submitted to you in writing following a site visit and should include the proposed new equipment, what work is required, and the full cost, including labor. Do not give your business to a

Make sure your installer has experience with high-efficiency systems.

company offering to give you an estimate over the phone without ever looking at the job to be done. You should expect a home evaluation including an inspection of your current system and a heat-load calculation. As explained below, the size of the new system should be based only on this analysis, not on the current equipment. Using the calculation, the contractor should also be able to estimate what your energy bills are likely to be with the proposed system.

A good contractor should ask about any heating problems you have had with your old equipment and offer suggestions for addressing these issues. A number of tests can be performed on your existing system to diagnose problems and determine if a new system is recommended. These tests are detailed under Upgrading Your Existing Heating System (p. 87).

Finally, reliable contractors are professional. Their people are prompt and courteous. They should have a published office or shop address. An office or shop is an indication that the company has been in business and intends to remain in business.

For More Information:

To find a heating contractor you can trust, consult the list of contractors certified by North American Technician Excellence.
www.natex.org ■ (877) 420-NATE

The Geothermal Heat Pump Consortium lists contractors that specialize in ground-source heat pumps
www.geoexchange.org

■ **Installation and Quality Control**
A new heating system must be installed properly to ensure safe and efficient operation. This includes making sure that any new ductwork or piping is properly sealed and insulated. Where possible, heating equipment should be installed within the thermal envelope of the house to minimize heat losses to the outside.

For ground-source heat pumps, proper installation will have to depend on the conditions of your house and lot. The first step, as with all systems, is to find a well-qualified contractor who has installed many of these systems. This can either be an HVAC contractor who will sub-contract the in-ground heat exchanger installation to a driller or excavator, or a driller who will sub-contract the HVAC side of the project. You want one contractor to be responsible for oversight of the whole installation and its performance.

With oil-fired furnaces and boilers, the system should be tuned and a combustion efficiency test performed after installation. Make sure the combustion efficiency meets the specifications (a calculation to convert from combustion efficiency to AFUE may need to be made). Gas furnaces and boilers come pre-adjusted from the factory and don't require a combustion efficiency test, but a good contractor will perform a few tests as a matter of course, including a CO (carbon monoxide) check.

If you are installing a condensing furnace or boiler (typically at least 90% AFUE), make sure it is properly vented per the manufacturer's require-ments, and a condensate drain has been installed. The drain tube should run into a floor drain or sanitary sewer line. Condensing furnaces and boilers typically employ a side-wall PVC (plastic) vent instead of venting through the chimney. If a chimney is used, the PVC should extend to the top, so no galvanized metal is exposed to the slightly acidic flue gas. Most condensing furnaces and boilers have sealed combustion, which is safer and better for indoor air quality.

■ Fuel Options

If you're buying a new heating system, the second decision (after the contractor) is what type of fuel to use. If you have an existing system, it usually makes sense to stick with the fuel your old system used, because the rest of the system will already be in place. But sometimes it makes sense to switch fuels.

Comparing cost. Table 4.2 demonstrates the true cost of different fuels per unit of useful heat delivered to the space. As a general rule, it is more costly and harder on the environment to heat with electric resistance heat, which includes electric furnaces, baseboard heaters, and portable space heaters. Electric resistance converts incoming electric current directly into heat, which means on-site efficiency is very high and there is less pollution near the house. But, as illustrated in

Chapter 1, when the inefficiency of electricity generation by the power company and transmission losses are taken into account, it is actually pretty inefficient to heat with electric resistance. Roughly one-third of the heating value of the fuel burned in a power plant will be delivered as useful heat in your house — the remaining two-thirds is lost to generation and transmission inefficiencies.

Switching between oil and gas. Switching from oil to gas or vice-versa based on current prices in your area can be risky. Even though one may be cheaper today, we expect their prices (per unit of heat energy) to converge, because so much of industry can use either fuel, depending on short-run price projections. Usually, fuel oil (and propane) prices fluctuate more quickly than natural gas.

Using electricity. If you currently have electric resistance heat, switching to gas or oil will probably save you money. Electric heat pumps can be more expensive than oil or gas heat in most of the country, because operating costs are highly dependent on proper installation. Of course, in warm regions the use of a heat pump for cooling can help justify its use for heating through milder winters.

When to choose a heat pump. Heat pumps are far more energy efficient than electric furnaces, and they can be used both for heating and air conditioning. But before deciding to replace your present system with a heat pump, you should carefully look into whether it makes sense in your climate.

Because air-source heat pumps rely on the outside air as the heat source in the wintertime, the colder that air, the worse the energy performance (the lower the coefficient of performance, or COP). For this reason, air-source heat pumps make more sense in warmer climates, where summer cooling loads are considerable. As the industry matures, we expect to see equipment and installations that are better suited for northern climates, and some heat pumps "tuned" for cold climates are on the market now (see "State of the Art").

Because temperatures underground are nearly constant year-round — warmer than the outside air during the winter and cooler than the outside air during the summer — a ground-source heat pump can be

TABLE 4.2 Comparing Fuel Costs per unit of Delivered Heat

How to use this chart: This chart allows you to compare the cost of different fuels according to the actual amount of heat delivered, in million Btus. First, find the price you pay for the fuel you use (you can often derive it from your utility bill). Then see where your price falls on the scale of true cost, and compare this to what you might pay for other fuels. The highlighted prices indicate the range of current prices across the U.S., and the price in bold indicates the average. The comparison assumes you are looking at high-efficiency options.

True Cost — $/million Btu scale: 5 10 15 20 25 30 35 40 45 50 55 60

Electric: resistance (100% efficiency) — ¢/kWh: 1 2 3 4 5 6 7 8 9 **10** 11 12 13 14 15 16 17 18 19

Electric: air source heat pump (250% efficiency) — ¢/kWh: 2 4 6 8 **10** 12 14 16 18 20 22 24 26 28 30 32 34 36 38 40 42 44 46 48

Natural Gas (95% efficiency) — $/therm: 0.50 1.00 **1.50** 2.00 2.50 3.00 3.50 4.00 4.50 5.00 5.50

#2 Fuel Oil (90% efficiency) — $/gallon: 0.50 1.00 1.50 2.00 **2.50** 3.00 3.50 4.00 4.50 5.00 5.50 6.00 6.50 7.00

Propane (95% efficiency) — $/gallon: 0.50 1.00 1.50 **2.00** 2.50 3.00 3.50 4.00 4.50 5.00

Source: *Your Green Home*, Green Building Press, 2006

much more efficient than an air-source unit and appropriate for both warm and cold climates. However, they are less common than air-source heat pumps and usually more expensive to install.

Despite the high cost, your energy bills might be lowered enough with a ground-source heat pump to justify installing one, especially if you need to replace your water heater as well. Most ground-source heat pumps are installed with a "desuperheater" that uses waste heat to heat water for no added cost during both heating and cooling modes. In addition, since the heat exchange pipes are underground and protected from the elements, ground-source units require less maintenance than conventional heat pumps, and there is no outside noise typically associated with conventional heat pumps and air conditioners.

Did You Know?

Ground-source heat pumps are easily 25–45% more efficient than new air-source heat pumps.

Several utilities, in cooperation with government agencies and other groups, have begun marketing ground-source systems as a superior alternative to air-source heat pumps, especially for new housing, where the cost and disruption of burying heat exchange coils is not as great as in replacement installations.

■ Switching Distribution Systems

In some cases, you might want to change from one type of distribution system to another. If you have steam heat, for example, switching to hot water may make sense. Some two-pipe steam systems can be retrofit for hot-water heat. Discuss this possibility with the heating technicians you talk with. If you have a forced warm-air or hot-water system, though, it rarely makes sense to switch to something else. The cost of adding either new ducts or new piping will probably make the project prohibitively expensive.

In old houses that were not originally constructed with central cooling, many homeowners have opted to tear out the old radiators and install a central forced-air system that can provide both heating and cooling. This can be a smart upgrade, but it deserves careful consideration. If the radiators are meeting your current comfort needs, it may be more efficient and cost less to keep your heating and cooling systems separate, since water transports heat so effectively. On the other hand, if you could benefit from the heat distribution and air filtration advantages of a forced-air system, then installing one may be a good way to solve a number of comfort concerns in one step.

If you are considering replacing both your furnace and your water heater, selecting a combination system that provides both heat and hot water may be advantageous to you. These are discussed in more detail in Chapter 6, under Indirect Water Heaters and Advanced Heating Systems with Integrated Water Heaters.

■ Sizing the System

Before you actually start shopping for a new furnace or boiler, it pays to figure out how large a system you need. A system that is too large wastes fuel and money because it keeps cycling on and off. It only runs at peak efficiency for short periods of time and spends most of the time either warming up or cooling down. Many systems that were installed in the 1950s and 1960s are much too large and the equipment may not have been downsized with subsequent replacements. It is not uncommon for a heating system to be two or three times larger than necessary.

> **Did You Know?**
>
> To be most efficient, your heating system should run virtually continuously on the coldest day of the year to keep your house at 70°F.

Remember, how much heat you need depends on how big your house is and how well it keeps heat in. Never figure out how big a system you need by checking the size of the old system. If your heating contractor wants to do this, take your business elsewhere. A heat loss analysis is the only way to determine the proper size of a new heating system. (Steam systems are an exception: the boiler should be sized to the radiators.) A heat loss analysis can be done by your heating contractor or energy auditor; it should include measurements of wall, ceiling, floor, and window areas and account for insulation levels and weatherization features, including any energy improvements you have made. From the heat loss analysis, you need to know the peak hourly heating demand in Btu/hr on the coldest expected day of the year in your area.

Furnaces and boilers are rated according to their Btu/hr output or heating capacity. A new heating system should be sized no more than 25% over the peak hourly heating demand. For example, if your home's peak hourly heating demand is calculated to be 60,000 Btu/hr, you should select a heating system with a heating output between 60,000 and 72,000 Btu/hr. The minimum acceptable heat loss calculation is called an "ACCA Manual J." Contractors with the technical capability to properly install modern systems should routinely perform this calculation. Ask to see a copy of the heat loss and system sizing calculations so that you can make sure they were thoroughly done

■ Dependability

When shopping for a high-efficiency furnace or boiler, buy a system with a good warranty and a reputable company to back it up. Most quality furnaces have a limited lifetime warranty on the heat exchanger, and most hot-water boilers come with a 20-year warranty. The controls should have at least a one-year warranty. If the manufacturer is a relatively new company, ask the dealer about the company's reputation. Ask for the names of several homeowners in your area with their systems in place, and call to see if the homeowners are satisfied with performance and service.

■ Efficiency Recommendations

The efficiency levels you want to look for vary according to the type of system and fuel, as indicated in Table 4.3. If you live in a cold climate and your house is well sealed, it usually makes sense to invest the extra few hundred dollars for the highest efficiency system available. In the mildest climates with low annual heating costs, the extra investment required to go from 80% to 90–95% efficiency may be hard to justify. Also remember that if your duct work is not properly sealed, choosing a 95% over a 92% AFUE furnace is probably not worth the money. Invest in duct sealing instead.

High-efficiency oil- and gas-fired systems—over about 85% AFUE—are typically condensing models. By wasting less heat as hot exhaust gases, you keep more money in your pocket. Most condensing units, but not all, have sealed combustion. We recommend asking for units with sealed combustion, which brings in outside air for combustion.

TABLE 4.3 Target Heating System Efficiency Ratings

	Air Source Heat Pump* (HSPF)	Ground Source Heat Pump (COP)	Gas Furnace (AFUE)	Oil Furnace (AFUE)	Gas Boiler (AFUE)	Oil Boiler (AFUE)
Market Range	7.7–10	2.5–3.2	78–96%	78–95%	80–99%	80–90%
ENERGY STAR	Split System: 8.2 Single Package Unit: 8.0	Open Loop: 3.6 Closed Loop: 3.3 Direct Expansion (DX): 3.5	90%	83%	85%	85%
CEE Tier 2	8.5	N/A	92%	N/A		
CEE Tier 3	N/A	N/A	94%			

If you live in a mild climate for heating, ACEEE recommends purchasing products at the ENERGY STAR level. If you live in a cold climate, then consider the higher Consortium for Energy Effiicency's (CEE) "Tiers," which offer additional savings. In many cases, CEE-member utilities offer rebates for highly efficient equipment.

* Heat pumps also have cooling-mode efficiency requirements that you'll want to specify. These are explained in Chapter 5.

For More Information:

Visit the Consortium for Energy Efficiency (CEE) website to learn more or to access a directory of certified furnaces, boilers, and heat pumps. For heat pumps, look under "HVAC directory" and for furnaces look under "Gas Programs."
www.cee1.org

To identify the most efficient furnaces, boilers, and ground-source heat pumps go the ENERGY STAR website and search under products for "heating and cooling."
www.energystar.gov ■ (888) STAR-YES

The Gas Appliance Manufacturers Association (GAMA) publishes product directories for furnaces and boilers.
www.gamanet.org

■ High Electrical Efficiency

The amount of electricity drawn by a furnace to power its motors and blow air through the house is often overlooked, since we focus on the natural gas or fuel oil used to make the heat. However, the electric energy demand by the furnace can be considerable — over 1,200 kWh/year for some models, adding up to $100 or more to your yearly electric bill. This power consumption is not factored into the AFUE ratings, so it is another item for consideration when choosing a new furnace. How much electricity a furnace uses depends on the design and configuration of the unit, the efficiency of your ducts, and most importantly, the efficiency of the motors and fans.

The actual amount of electricity used by any furnace will vary with your local weather and house characteristics, such as the efficiency of your ducts. As noted above, the most important variable is the fan motor type. Most units today use "permanent split capacitor" motors, but premium units are often equipped with variable speed fan motors, also called "ECM," or "permanent magnet DC" motors. These motors are more efficient than conventional furnace fan motors because they only use a high fan speed when needed. Lower speeds are generally used for heating and air circulation, while the highest fan speed is reserved for air conditioning. Variable speed motors also allow finer control and some premium features such as enhanced dehumidification in the cooling season and quieter air circulation when neither the heating nor the air conditioning is on. These can be very cost-effective where the furnace or central air conditioner runs many hours, or where the homeowner uses the fan to circulate air through the house and filter, regardless of whether the system is calling for heating or cooling. Recent work suggests that actual savings are relatively small in regions with low heating loads (for example, central California).

Ask your contractor for information on the type of fan and estimated electricity use of the fan included in any system recommended for your home. The Gas Appliance Manufacturers Association (www.gamanet.org) notes electrically efficient models in its furnace listings.

Upgrading Your Existing Heating System

Even if you aren't about to go out and buy a state-of-the-art, high-efficiency heating system, it is still a good idea to hire a contractor to check out your system, perform regular tune-ups, and look for opportunities to realize substantial savings by boosting the efficiency and performance of your present system. In a few situations, it makes more economic sense to tune up or modify your existing system than to replace it.

■ Routine Professional Tune-ups

Oil-fired systems should be tuned up and cleaned every year, unless they operate on ultra-low sulfur fuel. Gas-fired systems should be checked every two years, and heat pumps every two or three years. Regular tune-ups should cut heating costs, and they also increase the lifetime of the system, reduce breakdowns and repair costs, and cut the amount of carbon monoxide, smoke, and other pollutants pumped into the atmosphere by fossil-fueled systems.

The company that sells oil usually has trained technicians who can test your furnace or boiler, clean it, and tune it for optimum efficiency. Independent contractors provide this service as well. A complete tune-up may cost $100 or so, and can reduce your heating bill from 3–10%. Some companies perform these services as part of a regular service contract. Check to make sure that all of the services listed below are included:

Combustion efficiency. Incomplete combustion of fuel and excessively high flue gas temperatures are the two main contributors to low efficiency. If the technician cannot get the combustion efficiency up to at least 75% after tuning it up, you should consider replacing the system (Note that the combustion efficiency is different from annual fuel utilization efficiency or AFUE. For older burners, the AFUE can be estimated by multiplying the combustion efficiency by 0.85. Thus, if the combustion efficiency is 75%, the AFUE is around 0.75 x 0.85 = 64%.)

The technician should measure the efficiency of your system both before and after tuning it up and provide you with a copy of the results. Combustion efficiency is determined based on one or more of the following tests: 1) flue temperature; 2) percent carbon dioxide or oxygen; 3) smoke number (oil); 4) carbon monoxide; and 5) draft.

■ *Flue temperature.* High flue gas temperatures mean that a lot of heat (and money) is being lost up the chimney. Typical flue temperatures are:

Gas 300–600°F

Gas (condensing system)   100–140°F

Oil   400–600°F

Oil (flame retention burner)   300–500°F

■ *Carbon dioxide.* Carbon dioxide is the primary end product of fossil fuel combustion. Too little carbon dioxide indicates incomplete combustion. For an oil burner, the CO_2 concentration should measure between 10 and 12%. For gas, it should be between 7 and 9%. If an oxygen reading is taken instead, it should be between 3 and 6% for oil systems, or between 5 and 7% for non-condensing gas systems.

■ *Smoke (oil only).* Smoke indicates lack of complete combustion and is usually not present in gas systems. For an oil system, on a scale of 0 to 10, the smoke number should be no higher than 1.

■ *Carbon monoxide (gas only).* Carbon monoxide (CO) indicates incomplete combustion. For safety reasons, recommended CO levels in the flue gas should not exceed 400 parts per million.

■ *Draft.* Correct draft promotes complete combustion and reduces net loss up the chimney. A pressure gauge measures the overfire draft through the combustion chamber, and the breach draft through the flue pipe. Overfire draft pressure should be between 0.01 and 0.02 inches of water, and breach draft should be between 0.02 and 0.04 inches higher than the overfire draft. If you have a sealed combustion or power vented unit, the draft test is not relevant. If you have a conventional unit and the pressure readings fall below these ranges, the furnace or boiler probably is near the end of its service life and has very low efficiency.

Cleaning. Oil system parts to be cleaned include the burner (nozzle, electrodes, filters), combustion chamber, oil line filter, and flue pipe. For all systems, clean the heat exchanger surfaces. Oil nozzles and filters are often replaced rather than cleaned. Sediment should be removed from the boiler and steam lines; corrosion inhibitors may be added to the boiler.

Adjustments. Air and fuel flow adjustments will be made based on the results of efficiency testing. The internal thermostat on the furnace or boiler (fan thermostat or aquastat) should be calibrated to turn on and off at the appropriate temperatures.

Pumps and Fans. Pumps and fans should be inspected and lubricated if necessary.

■ Modifications by Heating System Technicians

During a routine tune-up, a heating technician may recommend additional modifications to your heating system. All of these measures for boosting the efficiency of your furnace or boiler require a professional with the proper training and tools.

Duct sealing. In homes heated with forced warm-air, ducts should be inspected and sealed to ensure adequate airflow and eliminate loss of heated air. It is not uncommon for ducts to leak as much as 15-20% of the air passing through them. And, leaky ducts can bring additional dust and humidity into living spaces. Thorough duct sealing costs several hundred dollars, but can cut heating and cooling costs in many homes by 20%.

A contractor can test your ducts to determine the extent and location of leaks. Accessible ducts are often sealed using mastic, which is applied to the outside of duct joints and other leak sites. A new alternative to mastic is aerosol-based duct sealing ("Aeroseal"). A machine connected to the ductwork blows a latex aerosol throughout ducts to seal leaks from the inside. This system can reach leaks in hard to reach or inaccessible spaces and effectively seals leaks up to 1/4" in diameter.

Did You Know?

About 20% of space conditioning energy is wasted through duct losses. Field studies demonstrate that aerosol-based duct sealing can reduce duct leakage by 80% on average.

Reducing System Size. If you have an older oil-fired system, and you've added insulation, upgraded your windows, or tightened your house, chances are that your burner runs for only a fraction of the time, even in the coldest weather. When your heating system constantly turns on and off, its performance is like driving in stop-and-go traffic: you don't get very good mileage. A simple way to reduce this waste is by decreasing the rate at which oil is fed into the burner. With oil systems, the service technician can install a smaller nozzle, which costs just a few dollars and can cut fuel bills by 5–10%. Nozzles are sized according to fuel flow rates. The specification plate on your burner should include an acceptable flow range; an average range would be from 0.50 to 1.25 gallons per hour (GPH). Nozzle size should not be reduced more than 25–30% below the lowest firing range on the specification plate. Burners for steam systems should not be downsized.

Draft reduction (oil only). The draft test will determine whether excess heat is being lost up the chimney. This problem is particularly common in systems that were converted from coal to oil. If the draft is too high, your service technician should install a barometric damper in the flue pipe. This may cost from $20–$100, but can reduce fuel use by 5% or more. If a barometric damper is already there, it may simply need adjustment.

New oil burner installation. If you have an old, inefficient oil burner but are not ready to replace the whole thing, have a flame-retention burner installed. It will mix oil and air more thoroughly, operate with less air flow, and send less heat up the chimney. In addition, a flame-retention burner will block air flow through the burner when the system is not running, reducing heat loss up the chimney. Flame-retention burners cost $400–600, depending on whether a new combustion chamber and controls are needed. A properly sized flame-retention burner with reduced nozzle size should save 10–20%. You'll do even better, though, replacing the whole furnace or boiler with a state-of-the-art high-efficiency model.

Outdoor reset (hot-water boilers only). An aquastat regulates boiler temperatures, keeping the water within a prescribed temperature range, usually around 180°F. Unfortunately, it will keep the water just as hot even when there is little need for heat, such as during spring and fall months. A modulating aquastat (or outdoor reset) senses outdoor temperature and keeps the boiler water only as hot as needed. Brand-name aquastats sell for $150–350 and reduce fuel consumption by 5–10%. You can control an aquastat manually as well (see the section below on operating your system). Condensing boilers without these controls will not save energy, since they will operate at temperatures where no flue gas condensing occurs.

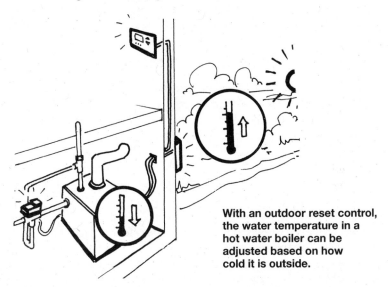

With an outdoor reset control, the water temperature in a hot water boiler can be adjusted based on how cold it is outside.

Time delay relay (hot water boilers only). Another strategy for controlling boiler water temperature is the time delay relay. When the room thermostat signals a need for heat, water heated earlier is circulated through the radiators without the boiler turning on. If circulation of warm water is not sufficient to heat the home within a specified time, the boiler burner fires to further heat the boiler water. With a time delay relay, circulation of lower temperature boiler water can provide adequate heating during milder weather. A time delay relay can be installed by a contractor for $50–75 and yield savings of 10%

Automatic vent damper. The automatic damper is a metal flap that closes off the flue when the burner shuts off. In practice, the electric vent damper is operated by the thermostat. In turn, opening the vent damper signals the boiler or furnace to turn on. This prevents the possibility of spilling combustion gases into the house if the damper malfunctions. Vent dampers cost from $125–400 installed and can cut fuel consumption by 3–15%. Savings are highest with steam boilers, large hot-water boilers, and warm-air furnaces that are located in heated spaces, where heated room air can escape up the chimney during the off cycle. If the heating system is located in an unheated basement or if it has a flame retention oil burner, savings will probably be less than 5%. If you have an older oil burner, converting it to a flame retention type is generally a better investment (see "New oil burner installation," above). Caution: vent dampers are not suitable for all gas heating systems. Ask your service technician whether a vent damper is appropriate for your system.

Adjustable radiator vents and valves. To reduce heat flow to unused rooms, valves on some hot-water radiators may be turned down. Valves on steam radiators should always be completely on or off, not in-between. An alternative for steam radiators is to install an adjustable air vent, typically costing about $10–15 at hardware and heating supply stores. These vents are screwed onto the radiator in place of existing vents, and they control how much steam gets into the radiator to heat it up.

You can get even greater control with either steam or hot water radiators by installing thermostatic radiator valves. These valves allow you to select the temperature of each room. When the designed temperature is reached, the valve shuts the radiator off. These valves cost $50–125 each installed, and can be a less expensive way to create separate heating zones, compared to repiping the whole house.

Programmable (or "clock") thermostats. Setting the thermostat manually works well but is inconvenient. More convenient is a programmable thermostat that will turn on the heat a half-hour before your alarm goes off in the morning. Some clock thermostats have several different set-back periods, helping you save energy when you go off to work and the kids leave for school. Programmable thermostats cost from about $40 to $150 for models that allow separate programming for each day of the week. Most will pay for themselves in about a year, if they are actually programmed properly and used consistently. This means several degrees of setback in winter and "set-up" in the air conditioning season.

Setting the thermostat back at night will save a lot of energy and money.

Tankless coil water heaters. If domestic hot water is provided by your heating system boiler with a tankless coil, then during the summer the boiler must operate constantly just to provide hot water for showers, washing dishes, etc. There are several ways to avoid this waste. The simplest but least convenient is to install a timer switch so that you can turn your whole heating system off at night and when you are away during the summer months. Another option is to install a stand-alone, gas-fired (or propane) water heater to use in the summer when your heating system is off. A third option is to install an indirect water heater that draws heat from the boiler (like a tankless coil), but stores the hot water so that the boiler does not need to run as frequently. This option is usually the most cost-effective alternative to tankless coils in cold climates.

A fourth solution may be to buy a solar water heater. In most areas of the country, solar water heaters can provide nearly 100% of summer-time hot water needs, thus complementing your wintertime boiler-produced hot water perfectly. Although this system requires a large capital investment, it is the best solution from an environmental stand-point, because solar energy produces virtually no pollutants, in contrast to fossil fuels and electricity. (See Chapter 6 for more information on water heating.) If you are building a new house, but not sure if you're ready to install solar water heating from the start, there are inexpensive steps you can take to support the installation at a later date. Plan for a south-facing roof and install "rough-in" plumbing; it will be cost-prohibitive to complete these retrofits later.

Insulate hot water pipes that run through unheated spaces. Be sure to use insulation that can withstand high temperatures.

■ Some Modifications You Can Do Yourself

You don't need to hire a contractor to take care of all the energy saving modifications. All of these modifications deal with heat distribution— getting the most heat from your furnace or boiler to the rooms in your house. These should be done regardless of the tested combustion efficiency of the heating system.

Pipe insulation. All hot-water and steam pipes passing through unheated areas should be wrapped with insulation. Use specially made foam or fiberglass pipe insulation, which can cost $0.30–0.80 per foot and saves about $0.50 per foot per year. Try to use insulation with a wall thickness of at least 3⁄4" for fiberglass, and 1⁄2" for foam. Do not wrap steam pipes with foam as it could melt. Older steam pipes may be wrapped with asbestos, which should not present a health hazard as long as it is well-sealed, not flaky, and not in a living space. If some of the white protective sheathing is missing, contact an asbestos abatement contractor.

Duct insulation. First, seal all duct joints and seams with mastic (a special paste) or UL-181b certified duct tape to keep hot air from leaking out of the ducts. Duct tape is that strong, silver or grey tape most of us know for its other applications, such as repairing broom handles and car parts. Unfortunately, non-UL-181 duct tape will dry up and lose its adhesion over time, especially in unheated spaces. Mastic, on the other hand, spreads easily and dries permanently, and is the preferred material for sealing joints and seams in metal ductwork. (See section on "Modifications by Heating System Technicians" for more on duct sealing).

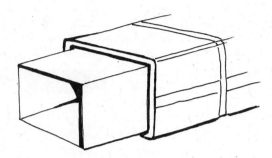

Duct installation is very important for ducts that pass through unheated spaces.

Secondly, all hot-air ducts passing through unheated spaces should be wrapped with insulation. You can use standard foil-faced fiberglass insulation, keeping the foil facing out and visible; vinyl-faced insulation made especially for ducts; or rigid foam insulation. R-8 is recommended in cold climates. Then, seal all joints or seams in the insulation. Use UL-181b duct tape with standard fiberglass batts and rigid foam; with vinyl-faced duct insulation, use duct tape or double-over and staple the seams.

Radiator reflectors. Radiators are designed to heat the living space, but they can lose a lot of heat into the exterior walls they are installed against. You can reduce this loss by placing reflectors between the wall and the radiator. You can make reflectors from foil-covered cardboard, available from many building supply stores. The reflector should be the same size or slightly larger than the radiator. The foil should be periodically cleaned for maximum heat reflection.

■ **Proper Operation and Maintenance**
How you operate heating system controls and maintain the equipment can have a large effect on your heating bills. People living in identical houses can have utility bills that vary widely, with some families paying 50% more than others. Maintenance should be performed regularly, about once a year.

Check the thermostat. In all heating systems, turn down thermostats in unused rooms and when you don't need the heat. In most homes, you can save about 2% of your heating bill for each degree that you lower the thermostat. Turning down the thermostat from 70°F to 65°F, for example, saves about 10% ($100 saved per $1,000 of heating cost). Programmable thermostats, which automatically adjust the temperature setting one or more times per day, are widely available (see previous section).

Did You Know?

Setting your thermostat back 10°F for eight hours at night can save about 7% ($70 saved per $1,000 of heating cost).

You might also be able to turn the thermostats down somewhat when you're in the room if you do a little buttoning up of the house (see Chapter 2). Turning down thermostats even when you are in a room doesn't mean being uncomfortable. In fact, you can actually be more comfortable at a lower temperature setting under the right conditions. Eliminating temperature stratification in a room where the floor is a lot colder than the ceiling will help the most, and getting rid of air infiltration and cold drafts is the best way to do it. Covering windows at night with blinds or drapes also helps, as does higher humidity. Buy some house plants or a humidifier, but don't add so much moisture that you start seeing condensation on your windows.

Maintaining Forced-Air Systems. Furnaces and heat pumps require regular maintenance to help keep air distribution working.

■ **Clean or replace air filters.** Standard 1" deep fiberglass filters on warm-air furnaces and heat pumps should be checked once a month during the heating season and cleaned or replaced as necessary. Dust blocks the air flow and forces the blower to work harder, which raises electric bills and can lead to blower failure. These filters cost about $1 apiece and can usually be purchased at hardware stores. Some more expensive filters, generally much deeper and pleated, can run for 6 – 12 months before replacement.

■ **Clean registers.** Warm-air registers (particularly the return registers) should be kept clean and should not be blocked by furniture, carpets, or drapes.

■ **Check duct dampers.** Often, air ducts have dampers (adjustable metal flaps) in them to control flow. Shut off or turn down dampers and registers that heat the basement. Other dampers and registers can be turned down or off to control heat flow to various rooms. Unused rooms should be kept cooler than occupied rooms to save energy. Don't close off too much airflow, though, because it may cause trouble for the fan.

Maintaining Hot Water and Steam Systems. When performing maintenance on hot water and steam heating systems, you may come in contact with dangerously hot water and steam. Use caution. If you're uncertain about how to do something, call a service technician (or your landlord, if you rent).

■ Keep baseboards and radiators clean and unrestricted by furniture, carpets, or drapes. Air needs to freely circulate through them from underneath. Also, do not cover tops of radiators.

■ Bleed trapped air from hot water radiators. Trapped air keeps radiators from performing properly. Use a radiator key to bleed air out of hot-water radiators once or twice a season. Hold a pan under the valve and open it until all the air has escaped and only water comes out. If you are not mechanically inclined, you may want to have the technician show you how to do it the first time.

■ Follow prescribed maintenance for steam heat systems, such as maintaining water level, removing sediment, and making sure air vents are working. Check with your heating system technician for specifics on these measures and use caution: steam boilers produce high-temperature steam under pressure.

■ Adjust the aquastat. This is the thermostat that regulates the temperature of the hot-water boiler. Normally, the aquastat keeps water in the boiler around 160–180°F. In milder weather, however, you don't need boiler water that hot. The aquastat can be set manually to 120–140°F, reducing fuel consumption by 5–10%. (If your boiler has a tankless coil for domestic hot water, you may not be able to turn the aquastat down this far.) The aquastat control is usually located in a metal box connected to the boiler. If you cannot locate it, ask your service technician for assistance. See the modifications section for information about modulating aquastats.

STATE OF THE ART HEATING

Radiant floor heat generally refers to systems that circulate warm water in tubes under the floor. This warms the floor, which in turn warms people using the room. It is highly controllable, considered efficient by its advocates, and is expensive to install. It also requires a very experienced system designer and installer, and limits carpet choices and other floor finishes: you don't want to "blanket" your heat source.

Contact the Radiant Panel Association:
www.radiantpanelassociation.org ▪ (800) 660-7187

Combined heat and power (CHP) or cogeneration for houses is being seriously studied in some countries. The basic premise is to use a small generator to meet some of the electric demand of the house, and recover the waste heat (typically more than 70% of the heating value of the fuel) to heat the house (hydronic or water-to-air systems) and make domestic hot water. These systems are not yet widely available. They are likely to have the best economics in houses with high heating bills because the house cannot be feasibly insulated, such as solid stone or brick homes.

Contact Keyspan Home Energy Services ▪ Climate Energy, LLC
www.keyspanservices.com   www.climate-energy.com

Cold-climate heat pumps are special designs optimized for winter use. They are supposed to minimize the use of resistive heat, generally by using multi-stage compression sections and sophisticated controls. These are currently offered by some manufacturers, and are in field trials by several utilities.

Contact Hallowell International ▪ Nyle Special Products, LLC
www.gotohallowell.com   www.nyletherm.com

Chapter 5

Cooling Systems

More than three-quarters of all U.S. households have air conditioners. Energy consumption for home air conditioning accounts for almost 5% of all the electricity produced in the U.S. for all purposes, at a cost to homeowners of over $15 billion. That amount of electricity results in the release of roughly 140 million tons of carbon dioxide (CO_2) per year, or an average for homes with air conditioning of more than 1.5 tons per year. A switch to high-efficiency air conditioners and implementation of measures to reduce cooling loads in homes can reduce this energy use by 20–50%.

Did You Know?

Air conditioning use in the U.S. has more than doubled just since 1981. Ninety percent of new homes are being outfitted with central air conditioning systems, including over 75% of new homes in the Northeast.

As air conditioning becomes the norm in all regions of the country, it is important to realize that often the easiest and most affordable way to stay comfortable during the summer is to reduce your need for air conditioning. After discussing how to keep heat out of your house, this chapter covers different types of cooling systems, when to upgrade, how to select a new system, and what you can do to operate your existing system for maximum efficiency.

For More Information

To find the latest information on how to reduce the energy you use for cooling, visit www.aceee.org/consumerguide/cooling.

Reducing the Need for Air Conditioning

Information on how to reduce your cooling loads can be valuable to you whether you use an air conditioner or not. And even with the most efficient air conditioners, it makes a great deal of sense to do everything you can to reduce your need to use them. The following conservation measures are often so effective that houses in the northern third of the country and in mountainous regions can get by without air conditioning on all but the very hottest days. If you're planning to buy a new system, reducing the cooling load will save you a lot of money right away by letting you buy a smaller, less expensive system.

■ What Is Comfortable?

In looking at how air conditioning costs can be reduced, it helps to understand human comfort. The standard human comfort range for light clothing in the summer is between 72°F and 78°F and between 35% and 60% relative humidity, according to the American Society for Heating, Refrigerating and Air-Conditioning Engineers (ASHRAE). The comfort range can be extended to 82°F with modest air movement, as might be provided by ceiling fans, for example. Often the house can be kept within this range using little or no mechanical air conditioning.

■ Getting Rid of Unwanted Heat

There are three major sources of unwanted heat in your home during the summer: heat that conducts through your walls and ceiling from the outside air, waste heat that is given off by lights and appliances, and sunlight that shines through your windows. These are described below, along with techniques to reduce them.

Common heat sources in the house

Insulate and tighten your house. Whenever the outdoor temperature is higher than the indoor temperature, warm air will blow into the house through cracks. To reduce these gains, you can insulate and tighten your house. If you don't have wall insulation, have cellulose or fiberglass blown into the walls by a qualified insulation contractor. Tighten up your house to reduce infiltration. You might also want to install a radiant

Did You Know?

One of the most cost-effective energy conservation measures, for both heating and cooling, is to add extra ceiling insulation. Increase its depth to a full 12 inches.

barrier in the attic to cut down on summer heat gain. If properly installed, a radiant barrier can reduce cooling costs to some extent, particularly in the South. An energy auditor can help you decide which measures make the most sense for your house and how much they will cost (see Chapter 2).

Get Rid of Inefficient Appliances. Lighting, refrigerators, stoves, washers and dryers, dishwashers, and other household appliances are all sources of waste heat, raising the interior temperature of your house. The best solution is to buy energy-efficient products. Energy-efficient appliances and lights produce far less waste heat. Standard incandescent light bulbs, for example, emit 90% of their energy as heat — only 10% as light. Compact fluorescent lights, on the other hand, produce only a fraction of the heat (see Chapter 11). In some cases, you can delay heat-producing tasks, such as dishwashing, until the cooler evening hours. You might also consider relocating a freezer to the basement or garage, where it won't contribute its waste heat to your living space. And by planning your meals carefully, you can minimize use of the oven on the hottest days.

Make "cool" choices when roofing or painting. Lighter colors tend to reflect more solar radiation from your house, cutting down the amount of heat penetrating the roof and walls. This may seem like a minor tip, but using "cool" products may lower roof surface temperatures by 100°F and reduce your peak cooling demand by 10–15%. If you are replacing your roofing tiles or if your roof is black, look for shingles, coatings, and other roofing materials that have been rated for high "solar reflectance" and high "thermal emittance." If your neighborhood has certain codes for roof tiling or if you are otherwise concerned about the aesthetics of a light-colored roof, look for new products on the market that use familiar dark pigments that have been engineered to conduct less heat.

For More Information

ENERGY STAR qualifies highly-reflective roof products.
On the website, click on products and select "roof products" under the Home Envelope heading.

www.energystar.gov ■ (888) STAR-YES

Shade or improve windows. Sunlight shining in windows, particularly those on the east and west sides of the house, usually adds the largest amount of unwanted summertime heat. In addition, the sun heats up the roof and walls of the house, increasing heat conduction to the interior. With no shading of east and west windows, the interior temperature of a typical house could rise as much as 20°F on a hot day, either making your air conditioner work a lot harder, or making you a lot less comfortable.

The best way to eliminate solar heat gain is to provide effective shading. Use horizontal trellises above east- and west-facing windows, which collect more summer sun than the others. Plant tall trees (prune lower branches so as not impede summer breezes), or use awnings wider than the windows to provide shade. If you have a choice, place porches, sheds, and garages on east and west walls to provide further shading. Unless you have extensive areas of south glass, the south wall should not require summer shading because the summer sun is at too high an angle to cause much of a problem. Planting evergreen trees in front of south windows, however, will block beneficial winter gains.

If you're replacing windows, put in high-performance windows with low-e glazings that look perfectly clear yet block out a large percentage of unwanted heat gain (see Chapter 2). It makes sense to install windows with low solar heat gain coefficients (SHGC) on west and east walls, where heat gain from the sun is the greatest in the summer. Windows with high SHGC make more sense on the south walls, especially when you want to benefit from passive solar heating during the

There are many ways to get rid of unwanted heat.

winter months. Another way to reduce solar gain through windows is to install drapes with light-colored linings or operable blinds that will reflect sunlight back outside. Vertical blinds are particularly effective on east- and west-facing windows. Also choose lighter colors for roofs and walls to reflect sunlight and reduce conductive heat gain.

■ Improving Comfort with Air Movement and Ventilation

Ceiling fans. Ceiling fans work to cool by creating a low-level "wind chill" effect throughout a room. As long as indoor humidity isn't stifling, they can be quite effective. If your fan has a motor that can spin in either direction, you can use it to lower your energy costs all year round. During the summer, a counter-clockwise motion creates a breeze. In the winter, a clockwise motion creates an updraft that keeps warm air in the occupied space. The following table gives the necessary fan blade diameters for various size rooms. Fans should have multiple speed settings so that airflow can be reduced at lower temperatures.

TABLE 5.1 Ceiling Fan Size

Room area (sq ft)	Minimum fan diameter (inches)
100+	36
150+	42
225+	48
375+	52
400+	2 fans

Source: Don Abrams, Low Energy Cooling.

When shopping for a ceiling fan, look for the ENERGY STAR label. ENERGY STAR ceiling fans provide more effective fan cooling with less energy by using improved blade design, offering more control options, and including high efficiency light kits. If you already have a ceiling fan, make sure to adjust the blades when needed to minimize wobbling. You can also reduce the energy the fan uses by replacing the lighting component with one certified by ENERGY STAR. As an added bonus, these light kits require fewer bulb changes (see Chapter 11 for more). Also, remember that fans cool people — they don't actually reduce room temperature — so turn it off when you leave the room.

Window fans. Window fans for ventilation are a reasonable option if they are used properly. They should be located on the leeward (downwind) side of the house facing out. A window should be open in each room. Interior doors must remain open to allow airflow. Window fans can be noisy, especially on high settings, but they are inexpensive.

Whole-house fans. A permanent whole-house fan is a more convenient option than window fans, but will cost more than three or four window fans. Mounted in a hallway ceiling on the top floor, the fan sucks air from the house and blows it into the attic. If the attic is not closed off to the rest of the house, the best spot for a whole-house fan is right in the gable vent, the highest point in the house. The fan is usually covered on the bottom by a louvered vent. (To reduce heat loss through the fan during the wintertime, the entire assembly should be installed within an insulated, weather-stripped box with a removable or hinged lid.)

The fan should have at least two speeds, with the highest one capable of changing the entire volume of air in the house very quickly. Because the fan blows air into the attic, the attic must have sufficient outlet vents. The attic free vent area, including soffit vents, ridge vents, and gable-end vents, should be twice the free vent area of the fan opening. (Free vent area is a measure of the area of the vent opening minus the area blocked by screening and louvers.)

Before turning on the fan, be sure to open several windows in various areas of the house. If just one or two windows are open, the airflow through them will be intolerably high. For safety reasons, the fan should have manual controls (either in addition to or instead of automatic controls). A fusible link, which automatically shuts the fan down in case of fire, should be included for safety (the controls should be installed by a licensed electrician). The fan should be installed carefully; a loose installation can cause vibrations and excessive noise. Units with low-speed motors (700 rpm or less) will usually be less noisy. Ask for a demonstration.

The whole-house fan can be turned on as soon as the outdoor temperature drops about three degrees below the indoor temperature, for example, in the evening. The fan speed should be adjusted according to how quickly you want to cool the house down. Mechanical ventilation uses far less electricity than mechanical air conditioning.

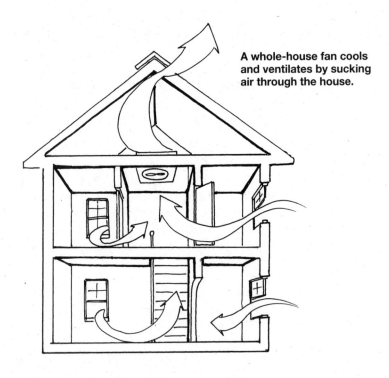

A whole-house fan cools and ventilates by sucking air through the house.

Air Conditioning Basics

Air conditioning, or cooling, is more complicated than heating. Instead of using energy to create heat, air conditioners use energy to take heat away. The most common air conditioning system uses a compressor cycle (similar to the one used by your refrigerator) to transfer heat from your house to the outdoors.

Picture your house as a refrigerator. There is a compressor on the outside filled with a special fluid called a refrigerant. This fluid can change back and forth between liquid and gas. As it changes, it absorbs or releases heat, so it is used to "carry" heat from one place to another, such as from the inside of the refrigerator to the outside. Simple right?

Well, no. And the process gets quite a bit more complicated with all the controls and valves involved. But its effect is remarkable. An air conditioner takes heat from a cooler place and dumps it in a warmer place, seemingly working against the laws of physics. What drives the process, of course, is electricity — quite a lot of it, in fact.

Types of Cooling Systems

■ Central Air Conditioners and Heat Pumps

Central air conditioners and heat pumps are designed to cool the entire house. In each system, a large compressor unit located outside drives the process; an indoor coil filled with refrigerant cools air that is then distributed throughout the house via ducts. Heat pumps are like central air conditioners, except that the cycle can be reversed and used for heating during the winter months. Heat pumps are described in more detail in Chapter 4. With a central air conditioner, the same duct system is used with a furnace for forced warm-air heating. In fact, the central air conditioner typically uses the furnace fan to distribute air to the ducts.

Central air conditioners and air source heat pumps operating in the cooling mode are rated according to their seasonal energy efficiency ratio (SEER), which is the seasonal cooling output in Btu divided by the seasonal energy input in watt-hours for an "average" U.S. climate. Many older central air conditioners have SEER ratings of only 6 or 7. The average central air conditioner sold in 1988 had a SEER of about 9; by

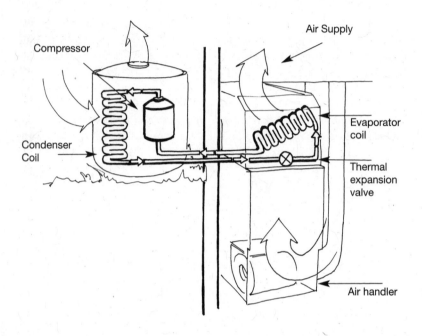

Air conditioners and heat pumps use the refrigerant cycle to transfer heat between an inside unit and an outside unit. Heat pumps differ from air conditioners only in the special valve that allows the cycle to reverse, providing either warm or cool air to the inside.

2002 it had risen to 11.1. The national efficiency standard for central air conditioners and air source heat pumps now requires a minimum SEER of 13, and to qualify for ENERGY STAR requires a SEER of 14 or higher. Central air conditioners also come with an energy efficiency ratio (EER) rating, which indicates performance at higher temperatures. ENERGY STAR-qualified models must meet an EER requirement of 11.5.

Similarly, cooling performance of ground source heat pumps is measured by the steady state EER. The ENERGY STAR program's minimum requirements for ground source heat pumps are 16.2 EER for open loop systems, 14.1 EER for closed loop systems, and 15 EER for direct expansion (DX) units.

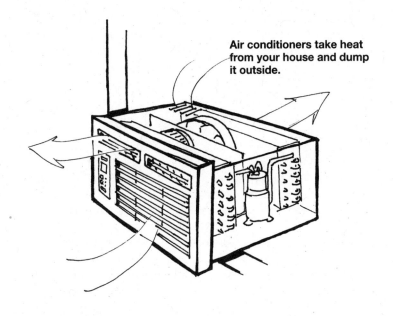

Air conditioners take heat from your house and dump it outside.

TABLE 5.2 Minimum Standards and ENERGY STAR Requirements for Room Air Conditioners

Capacity (Btu/Hr)	Federal Standard minimum EER	ENERGY STAR minimum EER
less than 6,000	9.7	10.7
6,000 to 7,999	9.7	10.7
8,000 to 13,999	9.8	10.8
14,000 to 19,999	9.7	10.7
20,000 or higher	8.5	9.4

■ Room Air Conditioners

Room air conditioners are available for mounting in windows or through walls, but in each case they work the same way, with the compressor located outside. Room air conditioners are sized to cool just one room, so a number of them may be required for a whole house. Individual units cost less to buy than central systems.

Room air conditioners are rated only by the EER, which is cooling output divided by power consumption. The higher the EER, the more efficient the air conditioner. Federal minimum efficiency standards for room air conditioners were revised in October, 2000; ENERGY STAR requirements exceed the federal standards by 10% or more. Table 5.2 lists requirements for units with louvered sides – the most common type.

■ Evaporative Coolers

Evaporative coolers, sometimes called swamp coolers, are less common than refrigerated air conditioners, but they are a practical alternative in very dry areas, such as the Southwest. They work by pulling fresh outside air through moist pads where the air is cooled by evaporation. The cooler air is then circulated through a house. This process is very similar to the experience of feeling cold when you get out of a swimming pool in the breeze. An evaporative cooler can lower the temperature of outside air by as much as 30 degrees. They can save as much as 75% on cooling costs during the summer because the only mechanical component that uses electricity is the fan. Plus, because the technology is simpler, it can also cost much less to purchase than a central air conditioner — often about half.

A direct evaporative cooler adds moisture to a house, which could be considered a benefit in very dry climates. An indirect evaporative cooler is a little different in that the evaporation of water takes place on one side of a heat exchanger. House air is forced across the other side of the heat exchanger where it cools off but does not pick up moisture. Both types begin to lose their effectiveness with increasing humidity, because humid air is less able to carry additional moisture.

For More Information

Leading manufacturers of evaporative coolers are listed at www.evapcooling.org/members.htm.

For evaporative coolers to do their job, they must be the right size. The cooling capacity of an evaporative cooler is measured not in the amount of heat it can remove (Btu), but in the fan pressure required to circulate the cool air throughout the house, in cubic feet per minute (cfm). A good rule is to figure the cubic square footage of your house and divide by 2. For example, a 1,500 square foot house with 8 foot-high ceilings would require a 6,000 cfm cooler.

■ Ductless Mini-Split Air Conditioners

Mini-split systems, very popular in other countries, can be an attractive retrofit option for room additions and for houses without ductwork, such as those using hydronic heat (see Chapter 4). Like conventional central air conditioners, mini splits use an outside compressor/condenser and indoor air handling units. The difference is that each room or zone to be cooled has its own air handler. Each indoor unit is connected to the outdoor unit via a conduit carrying the power, refrigerant and condensate lines. Indoor units are typically mounted on the wall or ceiling.

The major advantage of a ductless mini-split is its flexibility in cooling individual rooms or zones. By providing dedicated units to each space, it is easier to meet the varying comfort needs of different rooms. By avoiding the use of ductwork, ductless mini-splits also avoid energy losses associated with central forced-air systems.

The primary disadvantage of mini-splits is cost. They can cost 30% more than a typical central air conditioner of the same size. But, when considering the cost and energy losses associated with installing new ductwork for a central air conditioner, buying a ductless mini-split may not be such a bad deal, espcially considering the long-term energy savings. Talk with your contractor about what option would be most cost-effective for you.

Should I Replace My Existing System?

This can be a difficult question. If your central air conditioner is older than 15 years, it is likely to fail soon anyway. Do your research now to find the best high-efficiency replacement before you are forced to do so by necessity (a typical air conditioner will last about 15 years, and a typical heat pump will last 10 years).

If your air conditioner is 10–15 years old, the decision to replace it depends on its size, condition and performance. A ten-year-old central air conditioner probably has a SEER rating between 7 and 8. If your existing system is SEER 10 or below, even an upgrade to the lowest

Selecting a New Cooling System

To ensure a reliable, high efficiency system, you must first find a skilled contractor with experience in high-efficiency cooling systems. Walk through these options with your contractor as they may apply to your particular home. (See "Choosing a Cooling Contractor.")

Climate	Options	Recommendations
CENTRAL AIR CONDITIONING		
Hot/Dry	Central Air Conditioner or Heat Pump	SEER 15 and EER 11.6 Use a whole house fan to cool the house at night (p. 105).
	Evaporative Cooler	A good option with low operating cost (p. 110).
Hot/Humid	Central AC	SEER 15 and EER 11.6 Premium "dehumidistat" for good performance and comfort (p. 120).
	Heat Pump	SEER 15 High HSPF and quality installation for air and ground-source.
	Dehumidifier	ENERGY STAR model Adds operating cost, but may greatly increase comfort
Moderate	Central AC or Heat Pump	SEER 13 Install a whole-house fan to cool at night (p. 105).
All	Mini-split	Good aesthtics and efficiency. Practical retrofit.
	Window/Wall	ENERGY STAR model

efficiency new model is likely to save you 30% on cooling if the unit is installed correctly. With an ENERGY STAR model of SEER 15, you may cut your bills in half.

If you've made upgrades to your house as described in the first section, your air conditioner may well be oversized. If you haven't already, make these upgrades first before thinking about a new system. A home performance contractor can also help you decide whether it's time to upgrade. Chapter 2 provides more information on hiring home performance contractors and energy auditors.

You may not have to replace the entire air conditioner. Sometimes just the outdoor compressor component needs to be replaced, though it may be hard to find high-efficiency parts for low-efficiency models. If you're replacing just the compressor, make sure that the new outdoor unit is compatible with the indoor blower coil. The highly efficient outdoor unit will not reach its rated efficiency if it is not properly matched to the indoor unit. An air conditioning service technician should be able to help you match units effectively.

Selecting a New Cooling System

Choosing an air conditioner (or heat pump) is an important decision. Buying an inefficient model will lock you into high electric bills for years to come. Your decision will depend on your climate, and whether you are replacing an existing unit or installing an entirely new system. If you live in a hot, arid region, such as the Southwest, look into evaporative coolers or other new systems designed for hot, dry climates (see State of the Art Cooling). For the rest of the country, compressor-driven air conditioning systems are about the only choice, other than natural cooling, which is covered in the first section of this chapter. This discussion will focus on conventional compressor-type air conditioners (including heat pumps).

■ Choosing a Contractor

Room air conditioners can be purchased for as little as a few hundred dollars, while large central air conditioners and heat pumps can cost as much as $5,000. If any modifications need to be made in the ducting system for a central air conditioner, that can add substantially to the cost. It pays to shop around and get bids from a number of different contractors. There is substantially more variation among contractors in the quality of their work than there is among manufacturers, and a significant fraction of "equipment problems" can be tracked to improper sizing or installation.

Try not to use the lowest bid as the main criteria for selecting a contractor. Better contractors might charge more, but they may offer greater value. When evaluating bids, be sure to consider what you are getting for the price. Does the system come with a warranty? If so, how long is it? Air conditioner warranties range from one year for complete parts and labor, to five years for the compressor. Some manufacturers are now offering ten-year warranties on the compressors. If an existing system is being replaced, will the old unit be hauled away? Does the air conditioning contractor offer a service plan, and is it affordable? Make sure the contractor has been in business for a while and is fully bonded and insured. If you are not familiar with the company, ask for some local references and follow them up.

A good bid should be submitted in writing following a site visit, should and include the proposed new equipment, what work is required, and the full cost, including labor. Do not give your business to a company offering to give you an estimate over the phone without ever looking at the job. You should expect a home evaluation, including an inspection of your current system and a cooling-load calculation (called an "ACCA Manual J"). As explained below, the size of the new system should be based only on this analysis, not on the current equipment. Using the calculation, the contractor should also be able to estimate what your energy bills are likely to be with the proposed system. A good contractor should ask about any problems you have had with your old equipment and offer suggestions for addressing them. Finally, reliable

Make sure your installer has experience with high-efficiency systems.

contractors are professional. Their people are prompt and courteous. They should have a published office or shop address. An office or shop is an indication that the company has been in business and intends to remain in business.

■ Installation Check-List

To maximize efficiency, the outside part of a central air conditioner — the compressor — should be located in a cool, shaded place. The best place is usually on the north side of the house under a canopy of trees or tall shrubs. However, it shouldn't be choked by vegetation; the compressor needs unimpeded air flow around it to dump waste heat effectively. Never place the compressor on the roof or on the east or west side unless it is completely shielded from the summer sun, because sunlight shining on it will heat it up and reduce its efficiency at dumping heat.

Also, the compressor may be somewhat noisy. Try to keep it some distance from a patio or bedroom window. If you're concerned about noise, ask to see (and hear) one in operation before buying it.

With heat pumps, location of the outside unit is more complicated. Because a heat pump is used for both cooling and heating, it usually makes more sense to locate the compressor on the south side — especially in colder climates, and shade it with a sunscreen or tall annuals during the summer. In the winter, when the compressor is trying to extract heat, a southern location will allow it to absorb solar energy, which will boost its heating efficiency.

**If possible,
locate room air conditioners
on a north wall or
a wall that is shaded.**

**The outside compressor for a
central air conditioner should
be shaded from direct sun.**

The location of room air conditioners is constrained by available walls and windows. To perform most efficiently, these units should be out of direct sunlight; if you have a choice of walls, the north is best and the south is second best; avoid east or west walls if at all possible. If you are shopping for a room air conditioner, ask the sales staff if you can listen to different models in operation.

■ Choosing the Right Type of Air Conditioner

Whether you should choose central or room air conditioning depends in large part on your climate and cooling loads. In small homes and those with modest cooling needs, room air conditioners often make the most sense. In fact, in a small, highly insulated house, even the smallest central air conditioner may be too large. If you are considering room air conditioners, you will need to decide between units that mount in the window and those that are built into the wall. Wall-mounted units are often a better choice, both for aesthetic and practical reasons, though they will cost more to install, because an opening has to be cut through the wall. Window air conditioners are harder to seal, they block views and light, and they prevent the use of the window for natural or forced ventilation.

Central air conditioners have a number of advantages. They are out of the way, quiet, and convenient. If you already have a forced-air heating system, you may be able to tie into the existing duct work. Whether or not your existing ducting will work for air conditioning depends on its size and your relative heating and cooling loads. Ask your air conditioning service technician. Plus, central air conditioners are more efficient (see Efficiency Recommendation).

Heat pumps, though more expensive, provide heat in addition to air conditioning all in one unit. If you already have a satisfactory gas or oil heating system and have decided to add air conditioning, it usually doesn't make sense to consider a heat pump, because even a high-efficiency heat pump will probably be more expensive to operate than your gas or oil heating system (see the discussion on heat pumps in Chapter 4). There are also situations where heat pumps can be beneficial. If you currently have electric resistance heat and you live in a relatively warm climate (winter temperatures seldom dropping below 30°F), a heat pump may be a good choice.

If you're unsure about which type of air conditioner makes the most sense for your house, ask for opinions and bids from several local air conditioning installers.

■ Sizing the System

No matter what type of system you choose, make sure that it is sized properly. Most air conditioners are rated in Btu/hour, but central air conditioners and heat pumps may also list cooling capacity by the ton. One ton is equivalent to 12,000 Btu/hour. With air conditioning systems, equipment cost is much more proportional to size than it is with heating equipment. Don't let a salesperson convince you to buy an oversized system. In addition to the higher cost for an oversized system, it will run only for short periods, cycling on and off, which will increase electricity use and decrease the unit's overall efficiency. If it just runs for short periods of time, it also won't do as good a job dehumidifying the air (see Dehumidification).

Find a qualified air conditioning technician or energy auditor to determine your cooling load. Do not rely on simple rules-of-thumb by air conditioner salespeople, but insist on thorough analysis, including local climate information and calculations of heat gain through windows and walls. Insist on an "ACCA Manual J" load calculation to size the equipment, and do not allow oversizing relative to the load calculation.

Wait to size the system until after you've taken measures to reduce your cooling loads as described in the first section of this chapter. Make sure conservation efforts are taken into account when the technician is figuring out how large a system you need.

With heat pumps, proper sizing can be especially difficult, because the same unit is used for both cooling and heating. A heat pump sized for heating loads in a cold climate will be considerably oversized when it comes to cooling, and a heat pump that is sized for cooling loads in a warm climate will tend to be oversized when it comes to heating. If the heating load is larger than the cooling load, some heat pump sales-people will recommend sizing the heat pump for cooling and then adding enough electric resistance heat to make up the difference in the winter. In such a situation, it generally makes more sense to size the heat pump to provide all of the heating requirements in average winter conditions, even though it will mean a larger and somewhat more expensive model. In this situation, a 2-speed or modulating compres-sor may be a worthwhile investment. A good heat pump technician should be able to help you choose the best compromise between cooling and heating capacity.

■ Efficiency Recommendations

For central air conditioners and air-source heat pumps, look for a SEER of at least 15. Besides SEER, there are two other factors to check. The first is high temperature performance. Although it is not federally mandated, EER at 95°F is available. Look for at least EER 11.6. In addition, a small device called a "thermal expansion device" (TXV), preferably factory-installed, should be specified. The TXV improves high-temperature performance, and helps the unit deliver its rated efficiency even under adverse conditions (such as inaccurate or low refrigerant levels). The device costs much less than the service call or inefficiency it guards against. When shopping for room air conditioners, look for EER ratings of 11 or higher, if available.

High-efficiency units generally cost more, but in hot climates more efficient units pay for themselves over a few years through reduced elec-tricity bills. Central air conditioners are usually more efficient than room

TABLE 5.3 Target Efficiency Ratings for Central Cooling Equipment

	Central Air Conditioner	Air Source Heat Pump*	Ground Source Heat Pump*
Market Range	13–21 SEER 9–14 EER	13–17 SEER 9–13.5 EER	8.7 – 20.4 EER
ENERGY STAR	14 SEER 11.5 EER	14 SEER 11.5 EER	Open Loop: 16.2 EER Closed Loop: 14.1 EER Direct Expansion (DX): 15.0 EER
CEE Tier 2	15 SEER 12.5 EER	15 SEER 12.5 EER	N/A
CEE Advanced Tier 3	16 SEER 13 EER	16 SEER 13 EER	

If you live in a mild climate for cooling, ACEEE recommends purchasing products at the ENERGY STAR level. If you live in a hot climate, then consider the higher Consortium for Energy Effiicency's (CEE) "Tiers," which offer additional savings. In many cases, CEE-member utilities offer rebates for highly efficient equipment.

* Heat pumps also have heating-mode efficiency requirements that you'll want to specify. These are explained in Chapter 4.

For More Information

Visit the Consortium for Energy Efficiency (CEE) Website to learn more about their recommendations or to access a directory of certified central air conditioners and heat pumps. You can limit your search by indicating size, "CEE Tier," etc.
www.cee1.org

To identify the most efficient room air conditioners and ENERGY STAR-qualified central cooling systems, go to the ENERGY STAR website and look under products for "heating and cooling."

www.energystar.gov ■ (888) STAR-YES

air conditioners, and in general, larger capacity air conditioners have higher efficiency. However, don't buy a larger system than you need just because it has higher efficiency (see discussion on sizing above).

Other energy-saving features to look for include a fan-only switch, which will enable you to use the unit for nighttime ventilation (see discussion on ventilation below). A filter check light to remind you to check the filter after a pre-determined number of operating hours is helpful. With central air conditioners and heat pumps, a variable speed fan will allow efficient air circulation for ventilation and filtering.

■ Dehumidification

Air conditioners remove moisture from the home as room air is forced past cold coils. Water vapor from the air condenses out on the coils the same way moisture from the air condenses on a glass of ice water on a hot, humid day. This water exits through a condensate drain.

Lowering the humidity in this way is both good and bad. You feel more comfortable at lower humidity levels, so the dehumidification contributes to cooling. But when water vapor condenses into liquid, it releases stored heat, reducing the apparent efficiency of the air conditioner. One of the ways manufacturers can boost the rated efficiencies of air conditioners is by keeping the condenser coils some-what warmer, thus reducing condensation. Some new air conditioners may not dehumidify air as effectively. This could be a problem, especially in humid climates.

High-efficiency air conditioning systems can get around the dehumidi-fication problem by including variable-speed or multi-speed blowers. High-speed operation leads to high efficiencies but low dehumidifica-tion. Lower speeds reduce efficiency but increase dehumidification. Lower speeds are used during very humid weather; the rest of the time a more efficient higher speed is used.

If you live in a humid climate, look for air conditioner models that are effective at removing moisture. Although there is no industry standard for rating the effectiveness at removing moisture, most literature does list water removal in pints per hour, which will help you compare one model to another. If you're buying a very high-efficiency air conditioner, choose a model with a variable-speed fan to aid in dehumidification, even though you might have to pay more for it. Several manufacturers have models with a variable-speed blower controlled by a humidistat, automatically reducing fan speed at high humidity. Also try to keep moisture out of the house (see Operation and Maintenance below).

Upgrading Existing Air Conditioners

The compressor units of most air conditioners have an average lifetime of only 10 to 12 years. By carefully following proper maintenance procedures, a quality model may hold up twenty years, but don't expect the kind of lifetime you get with boilers and furnaces.

■ Operation and Maintenance

Using natural or forced ventilation at night, while keeping the house closed up tight during hot days, is less expensive than operating your air conditioner (see Reducing the Need for Air Conditioning, above). Use air conditioning only when ventilation is inadequate. Don't cool unoccupied rooms, but don't shut off too many registers with a central system either, or the increased system pressure may harm the compressor. If your room air conditioner has an outside air option, use it sparingly. It is far more economical to recirculate and cool the indoor air than to cool the hot outdoor air to comfortable temperatures. Always keep all doors and windows closed when operating an air conditioner. Do not operate a whole-house fan or window fans while using the air conditioner.

You will probably be comfortable with the thermostat set at about 78°F, but ceiling fans can increase your comfort range. You will save 3–5% on air conditioning costs for each degree that you raise the thermostat.

Did You Know?

Proper air conditioner refrigerant charge can improve system efficiency by 20%. Correction of airflow rates can improve efficiency a further 5–10%. Have your system tuned up every 2–3 years to ensure the most efficient operation.

You can also increase comfort at warmer temperatures by reducing humidity; use a bathroom exhaust fan when you shower, run your dishwasher during the cooler evening or early morning hours, don't dry firewood in your basement, and don't vent your clothes dryer inside. And don't forget to set the temperature back (to a warmer temperature) when you leave the house. Rather than relying on your memory and manually resetting the thermostat, consider installing a programmable thermostat. See the detailed discussion in Chapter 4.

Air conditioners and heat pumps need regular maintenance in order to perform at peak efficiency. Clean the air filters on room air conditioners monthly. They should never be allowed to get dirty enough to impede air flow, as this could cause damage to the unit. The condenser should be cleaned by a professional every other year, or even yearly in dusty locations.

Central air conditioning units should be inspected, cleaned, and tuned by a professional once every two to three years. This will extend the life of the unit and reduce electricity consumption. Check with your service technician about the proper maintenance schedule for your unit.

During service of your unit, its refrigerant may need recharging. It is important that it is charged correctly. A 20% undercharged system can operate at 20% lower efficiency, and field studies show that the majority of central air conditioners are undercharged enough to affect performance. However, an overcharged system not only reduces operating efficiency, it can cause damage to the unit and reduce the lifetime of the system. Also, because refrigerants damage the ozone layer, federal law requires that the refrigerant be recovered and recycled.

The service technician should also measure airflow over the indoor coil. Recent studies show that inadequate airflow is a common problem and average airflow rates tend to degrade over time due to poor maintenance. Correction of airflow rates can improve efficiency by 5–10%.

Even if an air conditioner or heat pump is installed and maintained with adequate airflow and the appropriate level of refrigerant, the unit will not operate efficiently if the duct system is in poor condition. Duct sealing can reduce cooling energy use by 10–15%. (See Modifications by

Heating Service Technicians in Chapter 4 for more on duct sealing.) Power to a central unit should be shut off when the cooling season ends; otherwise the heating elements in the unit could consume energy all winter long. Flip the circuit breaker to turn it off if the unit doesn't have a separate switch. Turn the power back on at least one day before starting up the unit in order to prevent damage to the compressor.

STATE OF THE ART COOLING

Night Breeze is a new home climate control technology designed to save energy in hot, dry climates. It is essentially a powered whole house fan, air conditioner and indirect water heater integrated under one control system. In the summer, the system draws in as much cool outdoor air as possible to meet cooling needs — the air conditioner only kicks on if absolutely necessary. In the winter, a water-to-air heat exchanger extending from the water heater supplies warm air to the system.

Contact: Davis Energy Group
www.davisenergy.com/nb_page.html

Also appropriate for dry climates, the Coolerado Cooler is an evaporative cooling technology that is 100% indirect. It can offer four to six tons of cooling with an energy consumption of 1,200 watts. Its energy efficiency ratio (EER) is 40 or higher, making it two to three times as efficient as the best conventional air conditioners.

Contact: Coolerado, LLC
www.coolerado.com   (303) 375-0878

Thermal Energy Storage is a technology that is best for simply shifting energy use from peak to off-peak hours. It works by storing energy in ice — at night, electricity is used to freeze water, and during the day, the ice can cool air that is circulated throughout the house. Most cost-effective for people who live in climates that cool off at night and pay more for "peak" electricity use (e.g., in California), this technology is now available for residential use.

Contact: Ice Energy, LLC
www.ice-energy.com   (949) 215-2465

Chapter 6

Water Heating

After heating or cooling, water heating is typically the largest energy user in the home. As homes have become more energy efficient over the past 20 years, the percentage of energy used for water heating has steadily increased. This chapter looks at the high-efficiency water heaters available and how you can reduce water heating costs with your present water heater. For various reasons, there is not an ENERGY STAR program to recognize efficient water heaters, although one is being considered. Therefore, it is important to learn more about water heating so that you can make a good choice among competing technologies and approaches.

For More Information

To get the latest information on reducing the energy
you use for water heating, go to
www.aceee.org/consumerguide/waterheating.htm.

Think About Replacement Now

If you're like most people, you're unlikely to go out looking for a water heater until your existing one fails. That will happen at the worst possible time — like just after guests arrive for a week-long visit. You'll have to rush out and put in whatever is available, without taking the time to look for a water heater that best fits your needs and offers a high level of energy efficiency.

A much better approach is to do some research now. Explore the options and decide what type of water heater you want — gas or electric, storage or demand, stand-alone or integrated with your heating system, etc. Figure out the proper size for your household, not just in terms of gallon capacity, but first-hour rating as well (see Sizing Your Water Heater).

If possible, replace your existing water heater before it fails. Most water heaters have a lifespan of 10–15 years. If yours is up there in age, have your plumber take a look at it and advise you on how much useful life it has left. If it's in bad shape, replace it now before it starts leaking or the burner stops working. In fact, it often makes sense to replace an inefficient water heater even if it's in good shape. The energy savings alone could pay for the new water heater after just a few years, and you'll be happy knowing that you are dumping fewer pollutants into the air and less money down the drain.

Fuel Options

If you are looking to replace your water heater, first determine what type of fuel makes the most sense. If you currently have an electric water heater and natural gas is available in your area, a switch might save you money. Oil- and propane-fired water heaters also tend to cost less to operate than electric models, although the upfront costs can be substantially higher, particularly for oil-fired models.

Before you rule out electricity, though, check with your utility company. It may offer special off-peak rates that make electricity a more attractive option. With off-peak electricity for water heating, the utility company puts in a separate meter with a timer in it. You can only draw electricity through that meter during off-peak periods, when the utility company has more capacity than it needs and is willing to sell it less expensively.

Selecting a New Water Heater

Whether you're replacing a worn-out existing water heater or looking for the best model for a new house you're building, choose carefully. Look for a water heater that satisfies your hot water needs and uses as little energy as possible. Often you can substantially reduce your hot water needs through water conservation efforts (see Conserve Water).

■ Storage Water Heaters

Storage water heaters are by far the most common type of water heater in use in the U.S. today. Ranging in size from 20 to 80 gallons (or larger) and fueled by electricity, natural gas, propane, or oil, storage water heaters work by heating water in an insulated tank. When you turn on the hot water tap, hot water is pulled out of the top of the water heater and cold water flows into the bottom to replace it. The hot water is always there, ready for use. Because heat is lost through the walls of the storage tank (standby heat losses) and in the pipes after you've turned the faucet off (distribution losses), energy can be consumed even when no hot water is being used. New energy-efficient storage water heaters contain higher levels of insulation around the tank to reduce this standby heat loss. As for distribution losses — a problem common to all types of water heaters — look in the section on Upgrading Your Existing Water Heater for tips.

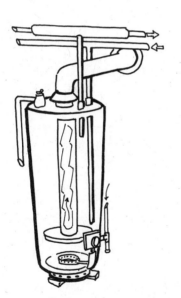

Storage or "tank-type" water heater.

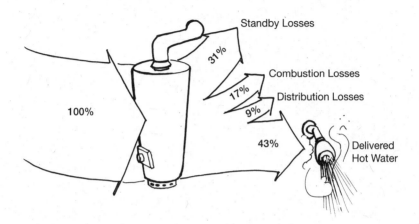

In a conventional gas storage water heater, less than 50% of the fuel energy reaches the point of use. New efficient water heaters can help reduce this excess heat loss.

Efficiency and tank size. The energy efficiency of a storage water heater is indicated by its energy factor (EF), an overall measure of efficiency based on the assumed use of 64 gallons of hot water per day, regardless of tank size. The first national appliance efficiency standards for water heaters took effect in 1990. Updated standards, effective in January, 2004, are summarized in Table 6.1.

All other things being equal, the smaller the water heater tank, the higher the efficiency rating. Compared to small tanks, large tanks have a greater surface area, which increases heat loss from the tank and decreases the energy efficiency somewhat, as mentioned above. If your utility company offers off-peak electric rates and you'd like to use them, you may need to buy a larger water heater to provide carry-over hot water for periods when electricity is not available under this "tariff."

The most efficient conventional gas-fired storage water heaters have energy factors of 0.65 and above, corresponding to estimated gas use of less than 250 therms/year. Condensing water heaters have energy factors as high as 0.86, and are gradually entering the market as smaller versions of commercial water heaters. For condensing water heaters with input capacity greater than 75,000 British thermal unit per hour (Btuh), look for the "thermal efficiency" rating rather than EF, with values of 0.90 and above.

The minimum efficiency of electric resistance storage water heaters is about 0.90 (depending on tank volume), and the best available are 0.95 EF. We do not recommend the use of electric resistance water heaters due to the high operating costs. A new electric water heater uses about 10 times more electricity than an average new refrigerator! Fortunately, heat pump water heaters using less than half as much electricity as conventional electric resistance water heaters are becoming commercially available. If you use electricity for water heating, consider installing a heat pump water heater. Look at how the costs compare over the life of a standard water heater in Comparing the True Cost of Water Heaters.

TABLE 6.1 Federal Requirements for Storage Water Heaters Minimum Energy Factor (EF)

Tank Size	Gas	Oil	Electric
30 gallons	0.61	0.53	0.93
40 gallons	0.59	0.51	0.92
50 gallons	0.58	0.50	0.90
60 gallons	0.56	0.48	0.89

Sealed combustion. For safety concerns as well as energy efficiency, look for gas-fired water heater units with sealed combustion or power venting. Sealed combustion (or "direct vent") is a two pipe system — one pipe brings outside air directly to the water heater; the second pipe exhausts combustion gases directly to the outside. This completely separates combustion air from house air. Power-vented units use a fan to pull (or push) air through the water heater — cooled combustion gases are vented to the outside, typically through a side-of-the-house vent. Power direct vent units combine a two-pipe system with a fan to assist in exhausting combustion gases.

In very tight houses, drawing combustion air from the house and passively venting flue gases up the chimney can sometimes result in lower air pressure inside the house (see Chapter 3). In turn, this can lead to "back-drafting," a situation when the air pressure inside is so low that the chimney airflow reverses, and dangerous combustion gases are drawn into the house.

▣ Demand Water Heaters

Demand or instantaneous water heaters do not have a storage tank. A gas burner or electric element heats water only when there is a demand. Hot water never runs out, but the flow rate (gallons of hot water per minute [gpm]) may be limited. By minimizing standby losses from the tank, energy consumption can be reduced by 10–15%. Before buying a demand water heater, though, be aware that they aren't appropriate for every situation.

The largest readily available gas-fired demand water heaters can supply about 5 gallons of hot water per minute with a temperature rise of 77°F (58° to 135°F, for example). 77°F is the basis for industry calculations. This would support two simultaneous showers, or a bit more if the hot water is "mixed down" with a lot of cold water. If you've installed a low-flow showerhead (see Conserve Water below) and won't need to do a load of laundry or dishes while someone is taking a shower, then 4–5 gpm might be fine. But if you have a couple of teenagers in the house, or if you need hot water for several tasks at the same time, a demand water heater might not be adequate.

Newer instantaneous gas water heaters modulate their output over a broad range; typical outputs might range from 15,000 to 180,000 Btuh—a 12:1 range in hot water output, depending on demand (washing hands versus washing a load of laundry). This is an improvement over earlier models. However, there are still some significant issues with tankless water heaters. First, they have a minimum water flow rate of 0.5 to 0.75 gallons/minute, and turn off or don't start at the lower flow rates used for hand washing and similar tasks. In some situations, this can lead to "cold water plugs" in the hot water supply lines, which leads to alternating delivery of hot and cold water — not regarded as a comfortable way to shower! They also require regular maintenance, and performance in hard water areas is not well studied.

Electric demand water heaters provide less hot water. A standard size model requires about 11 kW per gpm to achieve a 77° F temperature rise. Large units may require 40 to 60 amps at 220 volts, beyond the wired capacity of conventional houses. If you want to consider an electric unit, make sure your electrical wiring can handle the job before you make a purchase. However, a small electric demand unit might make good sense in an addition or remote area of the house, thereby eliminating the heat losses through the hot water pipes to that area. These losses often account for a large percentage of the energy wasted in water heating, regardless of the technology of the water heater. If you're using energy to heat water that has to make it a few hundred feet across the house, a small electric demand water heater placed under the sink to boost the temperature of incoming water locally may be a good idea.

Demand water heaters make the most sense in homes with one or two occupants, and in households with small and easily coordinated hot water requirements. Without modulating temperature control (above), you and your family may find yourselves unhappy with fluctuating water temperatures — particularly if you have your own well water system with varying pressure.

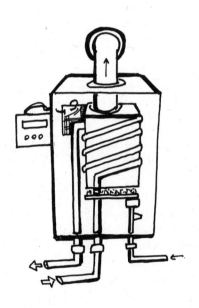

Demand, "tankless," or "instantaneous" water heater

With gas-fired demand water heaters, keep in mind that a pilot light can waste a lot of energy. In gas storage water heaters, energy from the pilot light is not all wasted because it heats the water in the tank. This is not the case with demand water heaters. A 500 Btu/hour pilot light can consume 20 therms of gas per year, offsetting some of the savings you achieve by eliminating standby losses of a storage water heater. To solve this problem, you can keep the pilot light off most of the time, and turn it on when you need hot water — a routine that should work fine in a vacation home, but not in a regular household.

A growing number of demand water heater models have electronic ignition, eliminating the need for a continuously burning pilot.

Manufacturers provide different specifications for demand water heaters: energy input (Btu/hour for gas, kilowatts [kW] for electric); temperature rise achievable at the rated flow; flow rate at the listed temperature rise; minimum flow rate required to fire the heating elements; availability of a modulating temperature control; and maximum water pressure. In comparing different models, be aware that you aren't always looking at direct comparisons, especially with temperature rise and flow rate. For example, while one model might list the flow rate at a 100°F temperature rise, another might list the flow rate at 70°. Until there are industry-standard ratings for temperature rise and flow rates, it will be difficult to compare the performance of products from different companies. Many manufacturers now publish energy factor ratings for these products and this information should make for easier comparisons. If you choose a tankless unit, look for one with EF 0.8 (gas) or EF 2.0 (electric).

What's a Tankless Coil?

Do not confuse a "tankless coil" with a tankless water heater or an indirect water heater. Tankless coils use the home's existing boiler — they are most common with older boilers — as the heat source for water heating with no storage tank. When hot water is drawn from the tap, the water circulates through a heat exchanger in the boiler. Tankless coils work great as long as the boiler is running regularly (during the winter months), but during summer, spring, and fall the boiler has to cycle on and off frequently, wasting a lot of energy. Instead, ACEEE recommends an "indirect" tank type storage water heater, or a free-standing storage water heater.

For More Information

The Gas Appliance Manufacturers Assocation (GAMA) lists high efficiency water heaters in the following categories on their Web site.

- Gas, Oil and Electric Storage
- Gas and Electric Tankless
- Electric Heat Pump
- Combination Space & Water Heating (indirect, integrated)

www.gamanet.org

■ Advanced Water Heating Systems

Some of the most efficient water heaters find creative sources of heat — such as other heating equipment, outside air, or the sun — to provide hot water with less fuel. These include heat pump water heaters, indirect water heaters, integrated space/water heaters, and solar water heaters.

Heat pump water heaters. More efficient than electric resistance models, heat pump water heaters use electricity to move heat from one place to another rather than generating the heat directly (see discussion on heat pumps in Chapters 4 and 5). The heat source is the outside air or air in the room where the unit is located. Refrigerant fluid and compressors are used to transfer heat into an insulated storage tank. While the efficiency is higher, so is the cost to purchase and maintain these units. Heat pump water heaters are available with built-in water tanks called integral units, or as add-ons to existing electric resistance hot water tanks. A heat pump water heater uses one-third to one-half as much electricity as a conventional electric resistance water heater. In warm climates they may do even better.

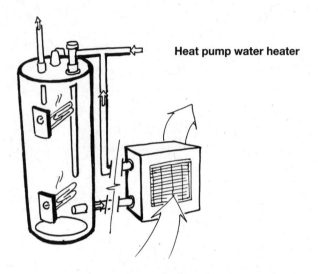

Heat pump water heater

Households using electric water heating and a heat pump for space conditioning can reduce water heating costs by installing a multi-function heat pump system. Fully integrated, single-unit systems are one option, or an existing heat pump and storage water heater can be retrofit with a specially designed add-on heat pump water heater module. Multi-function air- and ground-source heat pump systems are available.

Indirect water heaters. Indirect water heaters generally use the home's boiler as the heat source, circulating water from the boiler through a heat exchanger in a separate insulated tank. In the less common furnace-based systems, water in a heat exchanger coil circulates through the furnace to be heated, then through the water storage tank. Since hot water is stored in an insulated storage tank, the boiler or furnace does not have to turn on and off as frequently, improving its fuel economy. Electronic controls determine when water in the tank falls below a preset temperature and trigger the boiler or furnace to provide heat as long as needed. The more sophisticated of these systems rely on a heat purge cycle to circulate leftover heat remaining in the heat exchanger into the water storage tank after the boiler shuts down, thereby further improving overall system efficiency.

Indirect water heaters, when used in combination with new, high-efficiency boilers, are usually the least expensive way to provide hot water. These systems can be purchased in an integrated form, incorporating the boiler or furnace and water heater with controls, or as separate components. Gas, oil, and propane-fired systems are available. Any form of hydronic space heating — hydronic baseboards, radiators, or radiant heat — can be provided by boiler systems.

Integrated water heaters. If you're building a new home or upgrading your heating system at the same time you're choosing a new water heater, you might consider a combination water heater and space heating system. These systems, also called dual integrated appliances, put water heating and space heating functions in one package. Space heating is provided via warm-air distribution.

The efficiency of a combination water heater with integrated space heating is given by its combined annual efficiency, which is based on the AFUE of the space heating component and the energy factor of the water heating component. Look for combined annual efficiencies of 0.85 (85%) or higher.

Integrated gas heaters feature a powerful water heater, with space heating provided as the supplemental end-use. Heated water from the water heater tank passes through a heat exchanger in a central air handler to heat air. The fan blows this heated air into the ducts to heat the home. Many combination systems of this type are available at low initial costs, but space and water heating efficiency is often less than that of conventional systems. Models incorporating a high-efficiency condensing water heater, such as the Polaris® by American Water Heater, are exceptions. These models realize efficiency gains over traditional equipment, at 90% combined efficiency.

As you may have guessed, proper sizing of a dual integrated appliance is very important for economical performance, since both space heating and water heating are given from one "box." Product manufacturers should be able to identify your local distributors and contractors who are familiar with these products and their installation.

Indirect water heater

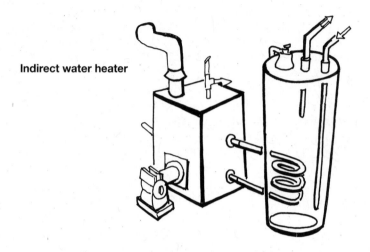

Solar water heaters. As the name implies, these use energy from the sun to heat water, or help heat water. Solar water heaters can be a great investment because they offer a virtually cost-free and renewable energy source for one of your home's top energy-users. But because the feasibility and benefits of a solar water heater will vary based on a number of variables, such as where you live, which way your roof is facing, and how many people live in your house, it takes some extra savvy to know what your costs and savings will be.

A solar water heater consists of a solar thermal collector attached to a south-facing sloped roof or wall, a well-insulated storage tank, and a fluid system that connects the two. It is usually preferable to use a two-tank system in which the solar water heater circulates water through the collectors and back into a separate tank that then "preheats" the conventional water heater. The distance between the collector and the tank, or the amount of finished space the loop must traverse in a retrofit installation, impacts the method and cost of installation.

Solar collectors can consist of an insulated glass box with a flat metal plate absorber that is painted black, or a set of parallel glass-encased metal tubes that absorb solar heat. Some collectors use a parabolic mirror to concentrate sunlight onto the tube. A fluid circulates between the solar collector and the tank either using a pump (active system) or natural convection (passive system). The most important difference among solar water heaters is the system's ability to resist freezing. If temperatures rarely go below freezing, water can circulate freely from the collector to the tank (open-loop system). But usually, it is safest to circulate another freeze-resistant fluid through the collector and trans-fer the heat to water using a heat exchanger in the tank (closed-loop system). An alternate strategy, which is simpler and often cheaper, is the "drainback" system, which is like an open system except it allows water to return to the tank as soon as the pump shuts off.

Solar water heaters are much less common than they were during the 1970s and early 1980s when they were supported by tax credits, but the units available today tend to be considerably less expensive and more reliable. The initial cost of a solar water heater is still much higher than other competing technologies, but if you can make the upfront invest-ment, it can save 50–75% of your water heating energy over the long term in most climates. For example, in New England it is possible to have a complete system that will provide one half of the hot water requirements for a typical family of four for around $5,000 installed. In most areas of the country, you will pay back the full price of the solar water heater over the course of its life. So it usually depends on whether you have the money and interest to make the upfront investment. See how a solar water heater stacks up against other technologies over its life cycle in Table 6.2.

Installation cost in your area may vary. Areas that receive sun consis-tently for 3 or more seasons will not only save more energy, but consumers are likely to have more products to choose from at lower costs. Plus, several Sunbelt states offer incentives for purchasing solar water heaters.

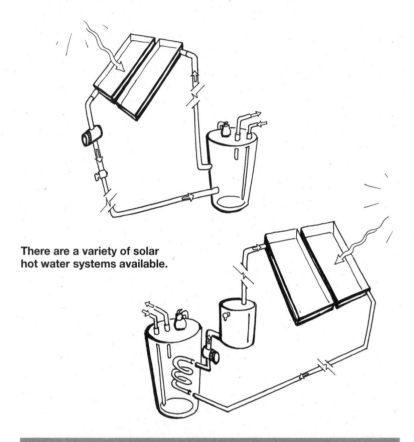

There are a variety of solar
hot water systems available.

For More Information

To see whether incentives for solar technologies are offered
in your state, contact the Solar Energy Industries Association.
www.seia.org ■ (202) 682-0556

For a list of manufacturers in your area, check with the
Solar Rating and Certification Corporation.
www.solar-rating.org ■ (321) 638-1537

Also check the Florida Solar Energy Center for a
list of rated high-quality solar water heaters.
www.fsec.ucf.edu ■ (321) 638-1000

If you are building a new house, it is to your advantage to have a south-sloping roof for easier installation of a solar water heater. At low cost, you can have the best place "pre-plumbed" or "roughed in" so "dry" pipes are in place if you later want to install a solar water heater.

If you have extra money to invest now and want to do more for the environment, a solar water heater can be a good choice in most areas today. But make sure you find a qualified installer who can properly design and size the back-up water heating system. Solar water heaters can be particularly effective if they are designed for three-season use, with your heating system providing hot water during the winter months.

Comparing the True Costs of Water Heaters

There are a number of important considerations when deciding what type of water heater you should buy: fuel type, efficiency, configuration (storage, demand, integrated), size, and cost. The information above covers most of these issues. Cost, however, needs some additional discussion. There are really two costs you need to look at: purchase price and operating cost.

It may be tempting when you're buying a water heater simply to look for a model that is inexpensive to buy, and ignore the operating cost. This course is penny-wise and pound-foolish. Often, the least expensive water heaters upfront are the most expensive to operate over the long run. Life-cycle costs, which take into account both the initial costs and operating costs of different water heaters, provide a much more accurate representation of the true costs of the water heater. Life-cycle costs for the most common types of water heaters under typical operating conditions are shown in Table 6.2.

From the table, we see that when both purchase and operating costs are taken into account, one of the least expensive systems to buy (conventional electric storage) is one of the most costly to operate over a 13-year period. An electric heat pump water heater, though expensive to purchase, has a much lower cost over the long term. A solar water heating system, which costs the most to buy, has the lowest yearly operating cost among electric systems.

TABLE 6.2 Life-Cycle Costs for 13-Year Operation of Different Types of Water Heaters

Water Heater Type	Efficiency	Cost[1]	Yearly Energy Cost[2]	Life (years)	Cost over 13 Years[3]
Gas Storage	0.60	$ 850	$350	13	$ 5,400
High-Efficiency Gas Storage	0.65	$ 1,025	$323	13	$ 5,200
Condensing Gas Storage	0.86	$ 2,000	$244	13	$ 5,200
Oil-Fired, Free-Standing	0.55	$ 1,400	$654	8	$11,300
Electric Storage	0.90	$ 750	$463	13	$ 6,800
High-Efficiency Electric Storage	0.95	$ 820	$439	13	$ 6,500
Tankless Gas (no pilot)	0.80	$ 1,600	$262	20	$ 5,000
Electric Heat Pump	2.20	$ 1,660	$190	13	$ 4,100
Solar-Assisted Electric	1.20	$ 4,800	$175	20	$ 7,100

[1] Costs are rough estimates, including installation, based on internal and other surveys.
[2] Based on hot water needs for typical family of four and energy costs of 9.5¢/kWh for electricity, $1.40/therm for gas, and $2.40/gallon for oil.
[3] Future operation costs are neither discounted nor adjusted for inflation.

Sizing Your Water Heater

To determine how big a storage water heater you need, you should first estimate your family's peak-hour demand. To do this, estimate what time of day (morning or evening) your family is likely to require the greatest amount of hot water. Then calculate the maximum expected hot water demand using Table 6.3. Note that this does not provide an estimate of your family's total daily use, only the peak hourly use. Also, the values in this table do not consider water conservation measures, like water-saving showerheads and faucet aerators that can reduce hot water use for each activity. If you have a new home, have upgraded to water-saving fixtures, or installed an ENERGY STAR clothes washer or dishwasher, your peak demand will be lower.

The ability of a water heater to meet peak demands for hot water is indicated by its first-hour rating. This rating accounts for the effects of tank size and how quickly cold water is heated. In some cases, a water heater with a small tank but powerful burner can have a higher first-hour rating than one with a large tank and less powerful burner. Ask appliance dealers for the first-hour ratings of appliances they sell or check the manufacturer's literature.

Buying too large a storage water heater will reduce energy performance by increasing the standby losses. If gas- or oil-fired, larger systems will also lose more heat up the flue.

TABLE 6.3 Peak Hourly Hot Water Demand

	Average gallons hot water per usage		Times used in hour		Gallons used
Showering	20	x	_____	=	_____
Bathing	20	x	_____	=	_____
Shaving	02	x	_____	=	_____
Washing hands and face	02	x	_____	=	_____
Hand dishwashing	04	x	_____	=	_____
Automatic dishwashing	10	x	_____	=	_____
Preparing food	05	x	_____	=	_____
Automatic clothes washing (warm or hot water)	32	x	_____	=	_____

For example, if your family's expected greatest hot water use is in the morning, the total might be:

3 showers	20	x	3	=	60 gallons/hr.
1 shave	02	x	1	=	02 gallons/hr.
Hand-wash dishes	04	x	1	=	04 gallons/hr.
Peak hour demand		=			66 gallons/hr.

Demand water heaters should be sized according to the required gpm flow rate and temperature rise during the winter required for your largest expected hot water fixture (usually a shower). (See Demand Water Heaters.)

With solar water heaters, you should discuss your requirements carefully with the solar water heating salesperson. You will need to size both the solar hot water system itself and the back-up electric or gas water heater. It generally makes the most sense to size a solar water heater to provide two-thirds to three-fourths of your total demand, and provide the rest with a back-up system.

Installing a Water Heater

Select an installation contractor carefully. Make sure that he or she has experience with the type of system you want to install. If the system is integrated with your heating system, have your heating contractor put in the water heater. To get a good value, ask for bids from several contractors and evaluate the bids carefully. Consider warranties, service, and reputation as well as the price.

Storage water heaters will lose less heat if they are located in a relatively warm area. Also try to minimize the length of piping runs to your kitchen and bathrooms. The best location is a centralized one, not too far from any of your hot water taps.

When the water heater is being installed, make sure heat traps or one-way valves are installed on both the hot and cold water lines to cut down on losses through the pipes. Without heat traps, hot water rises and cold water falls within the pipes, allowing heat from the water heater to be lost to the surroundings. Heat traps cost around $30 and will save $15–30 per year. Some new water heaters have built-in heat traps. If heat traps are not installed, you should insulate several feet of the cold water pipe closest to the water heater in addition to the hot water pipes. Even with heat traps, insulate the cold water line between the water tank and heat trap.

Upgrading Your Existing Water Heater

Even if you aren't going to buy a new water heater, you can save a lot of energy and money with your existing system by following a few simple suggestions.

■ Conserve Water

Your biggest opportunity for savings is to use less hot water. In addition to saving energy (and money), cutting your hot water use helps conserve precious water resources, a critical step in many parts of the country where supplies are scarce or water infrastructure is strained. Federal efficiency standards effective in 1994 set maximum water use requirements for showerheads and faucets at 2.5 gpm. Studies show that 1 in 4 households still use showerheads that exceed this flow rate. If you have older showerheads and faucets, consider replacing them now. Many high-quality, water-saving showerheads are available that provide impressive performance. Some models use just 1.6 gpm while offering multiple spray settings and even large water droplets. Of course, if your shower has multiple showerheads, you'll end up using a lot more water, even if each showerhead meets or exceeds the minimum standards.

If you aren't sure how much water your shower currently uses, you can find out with a bucket that holds at least a gallon and a watch with a second hand. Turn on the shower to the usual pressure you use, hold the bucket under it and time how many seconds it takes to fill to the one-gallon mark. If it takes less than 20 seconds, your flow rate is over 3 gpm, and your shower is a good candidate for a water-saving showerhead. Many older showerheads deliver 4–5 gpm.

> ## Did You Know?
>
> Water-saving showerheads and faucet aerators can cut hot water use in half. By installing new showerheads, a family of four can save 14,000 gallons of water a year and the energy required to heat it!

While faucets are also subject to the 2.5 gpm standard, there are opportunities to save even more water. Look for faucet aerators that deliver 1/2–1 gpm, especially if you tend to run the water when rinsing dishes. Some models are sold with a convenient shut-off valve at the aerator that allows you to temporarily turn off the water without changing the hot/cold mix. Sink aerators cost just a few dollars apiece. Repair leaky faucets right away. Even a relatively small hot water drip can dump a lot of energy and money down the drain. Also, refer to Chapters 9 and 10 for water-saving tips with dishwashers and clothes washers.

For More Information

Take a virtual tour to find water-saving opportunities in your home, calculate a water budget, and see the top tips for home water savings at the Water Saver Home site.
www.h2ouse.org

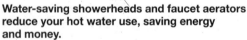

Water-saving showerheads and faucet aerators reduce your hot water use, saving energy and money.

■ Improve the Distribution System
Hot water distribution losses can be dramatic. Studies show that in a typical small house, 9% of the water-heating energy is lost in the distribution system. In a large house, distribution losses can be even greater. Insulating your hot water pipes will reduce losses as the hot water is flowing to your faucet and, more importantly, it will reduce standby losses when the tap is turned off and then back on within an hour or so.

A great deal of energy and water is also wasted waiting for hot water to reach the shower or tap. It is not uncommon for a home to waste thousands of gallons of water per year waiting for hot water to reach the tap. Homes with uninsulated pipes running beneath the foundation or with long distances between the water heater and bathrooms can experience a tremendous amount of heat loss. One common remedy is to install a hot-water recirculation loop to continuously pump hot water through the pipes ensuring that hot water is available almost instantly, even at fixtures furthest from the water heater. Continuous recirculation systems use a lot of energy to pump water through the system and lose a lot of energy from heated water traveling through the pipes. If a recirculation system is used, it is critical that the pipes are well insulated and that a timer is installed to turn the system off at night and at times when hot water use is minimal.

One superior alternative to a continuous recirculation system is a button-controlled system that delivers hot water only when needed by activating a small pump that quickly cycles the cool water standing in the hot water pipes back to the water heater, bringing hot water to the tap (these systems are discussed in greater detail under State of the Art Water Heating). If you find you often leave the shower unattended while you wait for the hot water, look for a device that cuts the shower flow to a trickle once hot water reaches the showerhead. This low-tech solution costs much less than recirculation, but it won't help if you find you often wait for hot water at the kitchen or bathroom sink.

▪ Improve Water Heater Efficiency

Insulate your existing water heater. Installing an insulating jacket on your existing water heater is widely-touted as one of the most effective do-it-yourself energy-saving projects. Newer water heaters come with fairly high insulation levels, practically eliminating the economic advantages of adding additional insulation. In fact, some manufacturers recommend against installing insulating jackets on their energy-efficient models. If you have an older water heater, consider upgrading to a new energy-efficient model for a great return on your investment. If replacement is out of the question for another year or two, an insulating jacket may make sense for the short term. Always follow directions carefully when installing an insulation jacket. Leave the thermostat(s) accessible. With conventional gas- and oil-fired water heaters, you need to be careful not to restrict the air inlet(s) at the bottom or the draft hood at the top.

Lower the water heater temperature. Keep your water heater thermostat set at the lowest temperature that provides you with sufficient hot water. For most households, 120°F water is fine (about midway between the "low" and "medium" setting). Each 10°F reduction in water temperature will generally save 3–5% on your water heating costs. If you have a dishwasher without a booster heater, you should probably keep the water temperature up to 140°F (the "medium" setting) or buy a new, more efficient dishwasher (see Chapter 9). Not only will lowering the thermostat save a lot of energy, but it will also increase the life of your water heater and reduce the risk of children scalding themselves with the hot water.

When you are going away on vacation, you can turn the thermostat down to the lowest possible setting, or turn the water heater off altogether for additional savings. With a gas water heater, make sure you know how to relight the pilot if you're going to turn it off while away.

STATE OF THE ART WATER HEATING

On-demand recirculation systems eliminate the energy, time, and water wasted when waiting for hot water to reach the faucet. The best systems consist of adding a connecting loop, a pump and a controller between the hot and cold water lines at the furthest fixture from the water heater. When activated by the push of a button, a pump rapidly circulates hot water to the fixture where it's called and the room temperature water in the pipes is returned to the water heater. Available products include the ACT Metlund® D'Mand® and Taco D'Mand® Systems.

www.gotohotwater.com ■ www.taco-hvac.com

Solar-powered circulators are available for solar water heaters – once you're using solar power for water heating, why not also supplement the electricity needed to run the system? These are available from several manufacturers. If you are considering a solar water heater, ask your installer about this option.

Drainwater recovery devices can be installed in the drainage lines beneath showers and other hot-water uses to recapture some of the heat returning to the pipe and transfer it back to the water heater. The most common example of this type of product is the GFX (gravity film exchange) by Doucette Industries.

Contact Doucette Industries:
www.doucetteindustries.com ■ 800-445-7511

Small-diameter and "Home run" piping systems are gaining popularity as an alternative to traditional copper piping. Instead of having a single, wide-diameter pipe that branches off to different end uses, individual polyethylene (PEX) tubes are run from a central manifold to each end use. Fewer fittings and elbows reduce friction and wait time.

www.ppfahome.org/pex

Chapter 7

Food Storage

The energy use of refrigerators and freezers has improved dramatically in the past 35 years, but they are still among the largest energy consumers in the home. A typical new refrigerator uses around 500 kWh per year — less than one-third the energy of a typical 1973 model — even though today's model is larger and has more features. This increase in efficiency has been achieved through more insulation, tighter door seals, larger coil surface area, better controls, and improved compressors and motors. Much of the increase in efficiency is due to national energy efficiency standards for new refrigerators. The current refrigerator standards took effect July 1, 2001, and new levels are slated to take effect in 2011 (pending legislative approval).

Buying a New Refrigerator

Because efficiency has improved so rapidly, if your refrigerator is old, needs repairs, or is nearing the end of its expected 15-year life, it may make good economic sense to replace it before it fails on you. For full-size refrigerators, ENERGY STAR units will use at least 15% less electricity than conventional units (20% less as of April, 2008), sometimes at no additional first cost. And

For More Information

Does it make sense to replace your existing fridge? Check out the Refrigerator Retirement Savings Calculator, click on appliances.
www.energystar.gov

some models even use 30% less electricity. We recommend that you consider models that are at least 20% better than the federal standard. Models with these levels of efficiency may qualify for rebates — check with your local utility.

In 2003, the ENERGY STAR program was expanded to cover compact refrigerators in addition to full-size models. Compact refrigerators less than 7.75 ft^3 must be 20% more efficient than the minimum federal standard. More on the economics of owning a second refrigerator is discussed below.

For More Information

To get the latest information on reducing the energy you use for food storage, go to www.aceee.org/consumerguide/food.htm.

To identify the most efficient products, go to the ENERGY STAR website and search under "products" for refrigerators and freezers.

www.energystar.gov ■ (888) STAR-YES

Configuration and Features

When shopping for a new refrigerator, it is important first to consider what style and features you want, and what the energy penalties can be. For example, side-by-side refrigerator/freezers use more energy than similarly sized models with the freezer on top, even if they both carry the ENERGY STAR. Icemakers and through-the-door ice also add to energy consumption. Unfortunately, it can be difficult to compare energy performance across different refrigerator types. The yellow EnergyGuide label and the ENERGY STAR program compare energy use among models with the same configuration within a narrow range of sizes. To compare models with different configuration or features, check the annual kWh per year or operating cost value displayed on the EnergyGuide label of each model of interest to you. Don't rely on the ENERGY STAR label alone when selecting a refrigerator — a top freezer model that does not qualify for ENERGY STAR may actually use less energy than an ENERGY STAR labeled side-by-side of comparable size and features.

TABLE 7.1 Comparison of Energy Use Across Refrigerator Types

Refrigerator Type	Total Capacity (ft³)	Fridge Capacity (ft³)	Freezer Capacity (ft³)	Energy Use (kWh/year)
Non-ENERGY STAR Models				
Top Freezer	21.9	15.5	6.4	529
Bottom Freezer	22.4	15.6	6.8	530
Side-by-Side	22.1	14.3	7.8	679
ENERGY STAR Models				
Top Freezer	21.7	15.2	6.5	422
Bottom Freezer	22.4	15.6	6.8	482
Side-by-Side	22.6	14.1	8.5	584

Note: All models compared here have icemakers; side-by-side models have through-the-door ice dispensers.

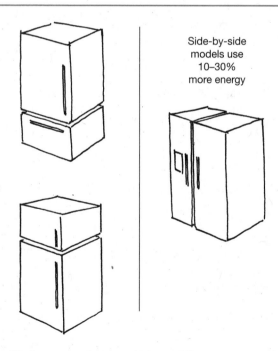

Side-by-side models use 10–30% more energy

The style and features you choose may impact the energy use of your refrigerator.

If maintained well (to minimize frost build-up, see section on boosting efficiency), manual defrost models use less electricity than automatic defrost models. However, they are not widely available. Built-in designer refrigerators may also consume more energy than store models, but are less wasteful than they used to be since the national appliance energy standards took effect.

Did You Know?

Features such as automatic icemakers and through-the-door ice and cold water dispensers will increase energy consumption by 10% to 20%.

Size

Size matters, too. Generally, the larger the unit, the greater the energy consumption. Too large a model will result in wasted space and energy; too small a model could mean extra trips to the supermarket. There is overlap between sizes, so it makes sense to compare. For example, you may find an efficient 18 ft^3 model that costs less to run than a 15 ft^3 model with similar features.

Keep in mind that the labeled capacity of the refrigerator may not correspond well to the actual usable space. A smaller top or bottom freezer model may actually have the same or more usable storage space than a larger side-by-side model, particularly in the refrigerator section.

Installing Your Refrigerator

If possible, locate refrigerators and freezers away from heat sources and direct sunlight. In the kitchen, try to keep your refrigerator away from the dishwasher and oven. Allow at least 1" space on each side of a refrigerator or freezer to allow good air circulation. Freezers can be installed in an attached garage or basement, which will boost energy performance somewhat during cooler months and reduce cooling loads in your house somewhat during the warmer months. However, don't put a refrigerator or freezer in a space that frequently goes below 45° — the refrigerant will not work properly.

Owning a Second Refrigerator

If you are thinking of buying a second refrigerator, or keeping your old one for extra storage when you get a new unit, you might want to reconsider. It is generally much less expensive to buy and operate one big refrigerator than two small ones. If the extra refrigerator is an old model, it's probably an energy guzzler. If you only need a second refrigerator a few days a year or to hold a few six-packs of beer, why spend an extra $100–150 per year in electricity? If you need a second refrigerator for convenient beverage and snack storage in a family room, consider an ENERGY STAR-rated compact refrigerator. If you have a freezer or second refrigerator that's nearly empty, turn it off. You'll do no harm to your refrigerator or freezer by turning it on and off periodically.

TABLE 7.2 Cost of Owning a Second Refrigerator

If you need more refrigerator space, consider the energy implications of different options.

Buying one large refrigerator	Buying a second compact fridge	Keeping your old fridge in the basement	
New ENERGY STAR Refrigerator (25 ft³)*	New ENERGY STAR Refrigerator (20 ft³) + New Mini-Fridge (5 ft³)	Your Old 1994 Refrigerator (25 ft³)	New ENERGY STAR Refrigerator (20 ft³)
505 kWh/year	762 kWh/year	737 kWh/year	432 kWh/year
$48 per year	$72 per year	$111 per year	

Assumes 9.5¢/kWh for electricity
* Bottom-freezer refrigerators are available in larger sizes than top-freezer models.

For More Information:

For information on environmentally responsible disposal of your old refrigerator and links to help you locate a recycler in your area, visit the ACEEE website www.aceee.org/consumerguide/disposal.htm

Buying a New Freezer

There are two basic freezer styles: upright (front loading) and chest (top loading). Chest freezers are 10–25% more efficient than uprights because they are better insulated and air doesn't spill out when the door is opened. A new standard for freezers also went into effect in July 2001. To help consumers identify the most efficient freezer, an ENERGY STAR labeling program for freezers was launched. To qualify for the label, full-size freezers must use at least 10% less electricity than the federal minimum standard, and compact freezers must use 20% less electricity. Manual defrost freezers are more common than automatic defrost models, and they tend to do a better job at storing food. (Automatic defrost freezers may dehydrate frozen foods, causing freezer burn.) Because a stand-alone freezer is opened less frequently than a refrigerator-freezer, frost will generally not build up as quickly as it might in a manual defrost refrigerator.

Boosting the Efficiency
of Your Existing Refrigerator or Freezer

From an energy standpoint, you will save the most by replacing your existing refrigerator or freezer with a new, more efficient model. If your current model is more than about 15 years old, it may be so inefficient that a new one would pay for itself in energy savings in just a few years. Unfortunately, that is often not practical. Refrigerators cost a lot. If your present one is working fine, it's hard to justify running out to buy a new one. So here are a number of ways to boost energy efficiency and performance of refrigerators and freezers.

■ Check Door Seals

Check the door seals or gaskets on your refrigerator/freezer. These can deteriorate over time, greatly increasing heat gain and decreasing energy performance. Put a dollar bill in the door as you close it; if it is not held firmly in place, the seals are probably defective. With newer magnetic door seals, this test may not work. Instead, put a bright flashlight inside the refrigerator and direct the light toward a section of the door seal. With the door closed and the room darkened, inspect for light through the crack. You will have to reposition the light as you move along the perimeter of the seal. Use a mirror to check the seal at the bottom of the door. If you don't see light, the seals should be in good shape. The dealer you purchased the refrigerator or freezer from should be able to install new seals. New seals aren't cheap, though. If the seals are bad, you might want to evaluate whether it's time to buy a new, high-efficiency model.

■ Check the Temperature

Check the temperature inside your refrigerator and freezer with an accurate thermometer. The refrigerator compartment should be kept between 36°F and 38°F, and the freezer compartment between 0°F and 5°F. If the temperature is outside these ranges, adjust the thermostat control. Keeping temperatures 10°F lower than these recommended levels can increase energy use by as much as 25%.

■ Move the Refrigerator to a Cooler Location

Take a look at where the refrigerator is located. If it's in the sunlight or next to your stove or dishwasher, it has to work harder to maintain cool temperatures. If you can move it to a cooler location, you'll boost energy performance. Also, make sure that air can freely circulate around the condenser coils. If that air flow is blocked, energy performance will drop (see Installing Your Refrigerator above).

■ Check Power-Saver Switch

Many refrigerators have small heaters built into the walls to prevent moisture from condensing on the outer surface — as if the refrigerator doesn't have to work hard enough already! On some units, this feature can be turned off with an energy-saver or power-saver switch. Unless you have noticeable condensation, keep this switch on the energy-saving setting.

■ Minimize Frost Build-Up

Manual defrost and partial automatic defrost refrigerators and freezers should be defrosted on a regular basis. The buildup of ice on the coils inside the unit means that the compressor has to run longer to maintain cold temperatures, wasting energy. If you live in a very hot, humid climate and don't use air conditioning, defrosting may be required quite frequently with a manual defrost model. After defrosting, you might be able to adjust the thermostat to a warmer setting, further saving energy.

■ Manage Your Food and Storage Space

There are several ways you can use your refrigerator differently so that it doesn't have to work as hard.

Let hot foods cool. Avoid putting hot foods directly in the refrigerator or freezer. Let them cool in the room first

Cover foods, especially liquids. Otherwise they will release moisture into the refrigerator compartment, increasing energy use.

Fill your freezer. A full freezer will perform better than a nearly empty freezer. If your freezer isn't full, fill plastic containers with water and freeze them. This can also help in the event of a power outage, when the ice you've made will help preserve your food longer.

Mark items. Label foods in the freezer for quick identification so that you don't have to stand there with the door open.

Chapter 8

Cooking

Choosing cooking appliances is a lot more complex and confusing today than it was 30 years ago. Along with the standard range with four top burners and an oven or two, we now have down-vented ranges with pop-out grills, fancy cooktops, microwave ovens, convection ovens, and a host of other smaller cooking appliances from high-tech toaster ovens and coffee makers to slow-cook crockpots, bread ovens, rice-cookers, and indoor electric grills.

While fuel choice and cooktop design are important considerations, the key to efficient cooking is understanding your cooking habits. Cooking patterns have changed considerably over the past few decades, with many Americans dining out more, cooking less, and using a microwave oven instead of a stove for preparing and heating many dishes. If you don't cook much, more efficient cooking appliances won't save much energy! On the other hand, these appliances tend to have long lives, so it is worth using efficiency as one guide when you purchase a new cooktop, stove, or oven.

Buying a New Range, Cooktop, or Oven

Reasons to consider purchasing a new range, cooktop, or oven may include a substantial change in cooking practices, failed or broken equipment, or replacement of an older gas range with a standing pilot light. If you are ready to buy a new appliance, first determine whether or not you would prefer to have your cooktop separate from your oven (or ovens). Standalone cooktops have grown in popularity in recent decades and have little to no impact on energy use. You must also make choices about size, fuel (gas or electric), and what cooktop element and oven type are most appropriate for your needs. Design options for cooktops and ovens are treated separately here, but the considerations apply equally to ranges, which combine the two in one unit.

For More Information

To get the latest information on reducing the energy you use for cooking, go to www.aceee.org/consumerguide/cooking.htm.

■ Fuel Choice: Gas or Electric?

Cooktops are widely available for either electric or gas cooking. Gas burners are often preferred by people who like to cook because gas offers a greater level of control in the speed of cooking. A downside to gas cooking appliances is that gas combustion products are introduced into the house and must be vented to the outside. Operate a ventilation fan that vents to the outside when using a gas cooktop or oven.

■ Gas Elements

There are three types of gas burners available: conventional burners with standing pilots, conventional burners with electric ignition (the most common), and sealed burners, where the burner is fused to the cooktop. In the U.S., standing pilots are not allowed on ranges or ovens with an electric cord. But they are allowed on cooktops. Standing pilots can more than double the annual energy consumption of your cooktop or range because they are continually burning a small amount of fuel, even when you're not cooking. When comparing sealed to unsealed burners, there is no measurable difference in cooking efficiency, although sealed burners are simpler and easier to clean.

■ Electric Elements

With electric cooktops, a number of different burner types are available, some of which offer energy savings and greater cooking control. The most common electric burners in this country are exposed coils, but you can also buy models with solid disk, radiant, halogen, or induction elements.

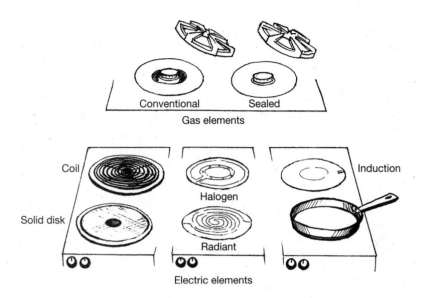

Conventional   Sealed
Gas elements

Coil

Solid disk

Halogen

Radiant

Induction

Electric elements

Some element designs offer greater efficiency and cooking control.

Radiant elements and halogen elements offer a level of efficiency just a bit higher than standard resistance coils. This is because they involve a heating source under a ceramic glass surface that provides for more rapid and even heat transfer. They are also easier to clean and generally more aesthetically appealing. Compared to standard resistance coils, these elements will likely add a few hundred dollars to the price of a new cooktop, although prices vary widely.

Solid disk elements are the least efficient electric cooktop design. They heat up more slowly than coil elements and generally use a higher wattage. It is especially important to use flat-bottomed cookware with these elements in order to minimize wasted energy.

Induction elements are the newest and most innovative type of electric cooktop in the North American market. Induction elements transfer electromagnetic energy directly to the pan where the heat is needed. As a result, they are very energy efficient — using less than half as much energy as standard electric coil elements while introducing less heat into the kitchen than conventional gas or electric designs. One catch is that they work only with ferrous metal cookware (cast iron, stainless steel, enameled iron, etc.). Aluminum or copper cookware will not work. When the pan is removed, there is almost no lingering heat on the cooktop. To date, induction elements have been available only with the highest-priced cooktops, but more and more manufacturers are adding them to their product lines at lower price points. Freestanding countertop units with one to four burners are also available.

■ Oven Type

In addition to standard electric and gas ovens, there are now convection ovens, microwave ovens, and combination models that work in one or more modes.

Conventional Ovens. With standard gas or electric ovens, self-cleaning models are more energy efficient because they have more insulation. But if you use the self-cleaning feature more than about once a month, you'll end up using more energy with the feature than you save from the extra insulation. If you're the type of cook who needs to peek into the oven all the time, buy a model with a window in the door and a light inside.

> **Did You Know?**
>
> Convection ovens use roughly 20% less energy than conventional ovens.

Convection Ovens. These can be more energy efficient than standard ovens because the heated air is continuously circulated around the food being cooked. You get more even heat distribution and, for many foods, temperatures and cooking time can be decreased. In some cases, you can cook multiple batches at the same time, cutting the additional energy consumption required for cooking each batch separately in a conventional oven. On average, you'll cut energy use by about 20%.

Microwave Ovens. Microwaves are very high-frequency radio waves. In these ovens, the energetic waves penetrate the food surface and heat water molecules inside. Energy consumption and cooking times for certain foods are greatly reduced, especially small portions and leftovers. Overall, energy use is reduced by about two-thirds. Because less heat is generated in the kitchen, you may also save on air conditioning costs during the summer. Some microwave ovens include sophisticated features to further boost energy efficiency and cooking performance, such as temperature probes, controls to turn off the microwave when food is cooked, and variable power settings. New "rapid-cook" ovens combining microwaves with other cooking technologies — notably halogen lights or convection — are designed to cut cooking time and improve the quality of foods compared to standard microwave preparation

Proper Ventilation for Gas Appliances

Proper ventilation for cooking appliances is very important for indoor air quality. The range hood should ventilate to the outside and not simply re-circulate and attempt to filter the cooking fumes. This is especially important with gas ranges. Whether you choose a conventional updraft hood or a downdraft hood, or vent can make a large difference in the amount of air that has to be heated or cooled to maintain comfort in the house: downdraft fans require several times more air than updraft models.

Also be careful about the size of the fan — too large a fan can waste energy and possibly even pose a safety risk. When a ventilation fan is operated, it depressurizes, or creates a slight vacuum in the house. To balance that pressure difference, outside air is sucked in through cracks in the walls and around windows (infiltration). That makes your heating system work harder, wasting energy. In some situations, that negative pressure can even prevent an oil or gas heating system from venting properly, causing backdrafting of dangerous combustion gases into the

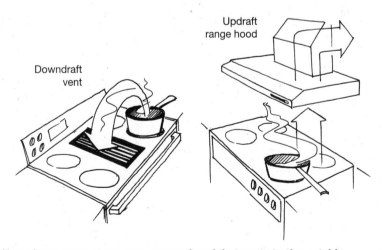

Updraft
range hood

Downdraft
vent

If you have a gas stove, use a range hood that vents to the outside.

house. This latter concern is especially serious with the large downdraft ventilation fans used with some cooktops and ranges. Ask about make-up air ducts for these ventilation systems. Clearly, downdraft range hoods require many extra steps to work safely because of the large air-flow required. This makes them both less efficient and more expensive.

Energy-Saving Tips for Cooking

Whether or not you plan to buy a new range or other cooking appli-ances, you can probably save a lot of energy just by modifying your cooking habits. With all the options you have for cooking a meal, decisions you make about what equipment to use and how to use it may save you the most energy and money.

Deciding how to cook a meal is not as easy as it used to be. Some options result in significant energy savings.

■ Appliance Choice and Proper Sizing
Some tools are better than others for cooking certain foods. Table 8.1 compares the energy use and cost for cooking the same meatloaf recipe using a range of methods. In this case, because meatloaf does not take up a large area, appliances that minimize the area that must be heated work better. On the other hand, the cheapest method, the microwave, may present a tradeoff between energy savings and food quality. The trick is to find the right balance, or an appliance explicitly

designed for a particular type of meal. For instance, for soups and stews that require long cooking periods, using a crockpot will save a substantial amount of energy.

TABLE 8.1 Energy Costs of Various Methods of Cooking

APPLIANCE	TEMP.	TIME	ENERGY	COST *
Electric oven	350°F	1 hr.	2.0 kWh	19¢
Convection oven (elec.)	325°	45 min.	1.4 kWh	13¢
Gas oven	350°	1 hr.	.11 therm	13¢
Frying pan	420°	1 hr.	.9 kWh	9¢
Toaster oven	425°	50 min.	.95 kWh	9¢
Crockpot	200°	7 hrs.	.7 kWh	7¢
Microwave oven	"High"	15 min.	.36 kWh	3¢

* Assumes 9.5¢/kWh for electricity and $1.20/therm for gas.

A large part of choosing the right tool is matching the size of the equipment to the size of the job. Full-size ovens are not very efficient when cooking small quantities of food. If you have two ovens, use the smaller one whenever you can. Some new range models are equipped with large and small oven compartments so it's easier to match the oven space to the size of your dish. If you make a lot of personal meals, it may make sense to invest in a toaster oven. Similarly, when using the cooktop, select the smallest pan necessary to do the job. Smaller pans require less energy to heat up.

Did You Know?

With electric cooktops, match the pan size to the element size. For example, a 6" pan on an 8" burner will waste over 40% of the heat produced by the burner.

■ Choosing the Right Cookware

Buying top-of-the-line pots and pans may be a low priority for your household. But it pays to understand the implications of certain designs and materials when you do decide to buy new cookware.

When cooking on electric burners, solid disk elements, and radiant elements under ceramic glass, cookware should rest evenly on the burner surface. The ideal pan has a slightly concave bottom — when it heats up, the metal expands and the bottom flattens out. An electric element is significantly less efficient if the pan does not have good contact with the element. For example, boiling water for pasta could use 50% more energy on a warped-bottom pan compared to a flat-bottom pan.

Certain materials also work better than others and usually result in more evenly cooked food. For instance, copper-bottom pans heat up faster than regular pans. In the oven, glass or ceramic pans are typically better than metal — you can turn down the temperature about 25°F and cook foods just as quickly.

For stove-top cooking, consider using a pressure-cooker. By building up steam pressure, it cooks at a higher temperature, reducing cooking time and energy use considerably.

■ Proper Maintenance

Believe it or not, cleaning your cooking appliances regularly will save energy. When burner pans — the metal pans under the burners that catch spills — become blackened from heavy use, they can absorb a lot of heat, reducing burner efficiency. Keep them clean and shiny and they'll be more effective at reflecting heat up to the cookware. Microwaves also work more efficiently when the inside surfaces are clean of food particles.

If you have a gas stovetop, make sure your burner is giving you a bluish flame. If the flame is yellow, the gas may not be burning efficiently. Have your gas company check it out.

■ How to Reduce your Cooking Time

Before You Start. Defrost frozen foods in the refrigerator before cooking so your oven or stovetop doesn't have to use its precious energy bringing your food to room temperature. With conventional ovens, keep preheat time to a minimum. Unless you're baking breads or pastries, you may not need to preheat the oven at all.

While You Cook. Food cooks more quickly and more efficiently in ovens when air can circulate freely. Don't lay foil on the racks. If possible, stagger pans on upper and lower racks to improve airflow if you're baking more than one pan at a time.

Try to avoid peeking into the oven a lot as you cook. Each time you open the door, a significant amount of heat escapes. Food takes longer to cook and you waste energy. Use your oven light and inspect through the window in the oven door instead.

With electric burners, you can turn off the burner just before the cooking is finished. The burner will continue radiating heat for a short while. This may also prevent overcooking. Another way to avoid overcooking in the oven is to use thermometers, especially for meat.

For Next Time. It doesn't take as much energy to reheat the food as it does to cook it. So cook double portions when using your oven, and refrigerate or freeze half for another meal. This will also save you preparation time!

If you have a self-cleaning oven, the best time to use the feature is just after you've cooked a meal — that way, the oven will still be hot and the cleaning feature will require less energy. Try not to use the self-cleaning feature too often, and operate the ventilation fan when it's on.

Chapter 9

Dishwashing

More than half of the energy used by a dishwasher goes towards heating the water. In fact, water heating accounts for approximately 60% of total energy use by dishwashers. Models that use less water also use less energy. Thanks to national efficiency standards, first effective in 1994, the energy and water consumed by dishwashers has dropped dramatically — while older models typically used 8 to 14 gallons of water, new models use only 3 to10 gallons per cycle. Even as dishwashers have become more miserly in their water use, they have made great strides in cleaning performance. With these advances come new standards for dishwashers, effective 2009 (legislation pending), that will restrict energy and water use even more.

Buying a New Dishwasher

There are a couple of quick ways to identify the most efficient dishwashers in the store or online. Today's ENERGY STAR dishwashers use 25–60% less energy per cycle than comparable non-ENERGY STAR models. For more information, look for the yellow EnergyGuide label for estimated energy use and a comparison to other models on the market. Although more energy-efficient models tend to use less water, there are variations in water use among ENERGY STAR models — some use half the water of others, saving hundreds of gallons of water each year. So to find the most water-efficient models, you must look beyond ENERGY STAR and EnergyGuide. Check the manufacturer's literature or contact your local water utility. In some states, electric and water utilities offer rebates for the purchase of models that are exceptionally efficient.

For More Information:

To get the latest information on reducing the energy you use for dishwashing, go to www.aceee.org/consumerguide/dishwashing.htm.

To identify the most efficient products, go to the ENERGY STAR website and search under "products" for dishwashers.
www.energystar.gov ■ 888-STAR-YES

In addition to looking for the ENERGY STAR label, keep in mind these important features affecting the energy and water use of dishwashers when shopping for a new model.

■ Energy-Saving Wash Cycles
Most dishwashers have several different wash cycle selections. If a load of dishes is only lightly soiled, a "light wash" cycle will save energy by using less water and operating for a shorter period of time. For example, the same dishwasher may use 5 gallons for a "light/china"

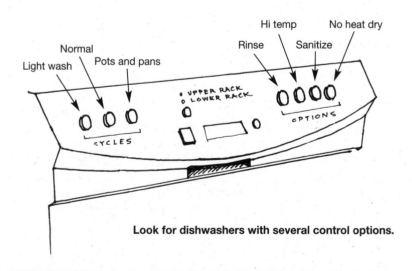

Look for dishwashers with several control options.

cycle, 7 gallons for a normal wash cycle and 9 gallons for the "pots/pans" cycle. Look at the manufacturer's literature for total water use with different cycles.

■ Soil Sensors vs. Standard Models

Many of the dishwashers on the market today use "soil sensor" technology to automatically adjust water use depending on how dirty the dishes are in each load. A new federal test procedure has been adopted to better estimate the energy consumption of soil-sensing dishwasher models. As a result, the data presented on updated EnergyGuide labels allows for more accurate comparisons among models with and without soil sensors. Highly efficient models of both types are available.

■ Energy-Saving "No-Heat" Dry

An electric heating element is generally used to dry dishes at the end of the final rinse cycle, consuming about 7% of dishwasher energy use. Most new dishwashers offer an energy-saving no-heat drying feature. At the end of the rinse cycle, if the feature is selected, room air is circulated through the dishwasher by fans, rather than using an electric heating element to bake the dishes dry.

■ Compact vs. Standard Capacity

Be aware that there are two dishwasher classifications: compact capacity and standard capacity. Compact models will use less energy, but they also hold fewer dishes. Compact models are usually 18 inches wide, compared to the 24-inch width of most standard capacity models. According to the U.S. Federal Trade Commission, to be considered "standard capacity," a dishwasher must be able to hold at least eight place settings of dishes. Some of the most efficient models are only 18 inches wide, yet they can accommodate eight full place settings of dishes, thanks to creative design of the dish racks. Larger models designed to accommodate up to 12 place settings are increasingly available and can save time and energy if you routinely have large loads. Just remember to wait for a full load before running the dishwasher. A compact dishwasher may actually result in more energy use if you have to run it more frequently. Another option is the "dish drawer" type of washer that allows you to wash either each drawer separately or wash a full load.

Installing a Dishwasher

When you install a dishwasher, try to position it away from the refrigerator. The dishwasher produces heat and will increase the energy consumption of your refrigerator. If it is a built-in dishwasher, you might be able to add extra insulation to the top, sides, and back when it's installed (check with the dealer first). This will both save energy and reduce noise levels. Use unfaced fiberglass batts.

Using a Dishwasher for Maximum Energy Savings

Whether you are buying a new dishwasher or using an existing one, you may be able to save a considerable amount of energy by changing the way you operate it. Below are tips on how to save energy when washing dishes.

■ Dishwashing vs. Hand-Washing

Which method uses less energy? Well, it depends on how old your dishwasher is, what settings you use, and how you would wash the dishes by hand. Studies are showing more and more that, when used to maximize energy-saving features, modern dishwashers can outperform all but the most frugal hand washers.

If you currently wash dishes by hand and fill sinks or plastic tubs with water, it's pretty easy to figure out whether you would use less water with a dishwasher. Simply measure how much water it takes to fill the wash and rinse containers. If you wash dishes by hand two or three times a day, you might be surprised to find out how much water you're currently using. Newer dishwashers use only 3 to 10 gallons per cycle.

■ Scrape, Don't Rinse

Studies show that most people pre-rinse dishes before loading them into the dishwasher. Modern dishwashers — certainly those purchased within the last 5–10 years — do a superb job of cleaning even heavily soiled dishes. Don't be tempted to pre-rinse dishes before loading, simply scrape off any food and empty liquids and let the dishwasher do the rest. This will save you time as well as water and energy. If you find you must rinse dishes first, get in the habit of using cold water.

■ When Filling the Dishwasher

Load dishes according to manufacturer's instructions. Completely fill the racks to optimize water and energy use, but allow proper water circulation for adequate cleaning.

Wash only full loads. The dishwasher uses the same amount of water whether it's half-full or completely full. Putting dishes in the dishwasher throughout the day and running it once in the evening will use less water and energy than washing the dishes by hand throughout the day. If you find that it takes a day or two to get a full load, use the rinse and hold feature common on newer models. This will prevent build up of dried-on food while saving time and water compared to pre-rinsing each item. The rinse feature typically uses only 1 to 2 gallons of water.

■ Use Energy-Saving Options

Pay attention to the cycle options on your dishwasher and select the cycle that requires the least amount of energy for the job.

Use the no-heat air-dry feature on your dishwasher if it has one. If you have an older dishwasher that doesn't include this feature, you can turn the dishwasher off after the final rinse cycle is completed and open the door to allow air drying. Using the no-heat dry feature or opening and air drying the dishes will increase the drying time, and it could lead to increased spotting, according to some in the industry. But try this method some time to see how well it works with your machine.

■ Turn Down the Water Heater Temperature

Since the early 1990s, most dishwashers in the U.S. have been sold with built-in heaters to boost water temperature to 140–145°F, the temperature recommended by manufacturers for optimum dishwashing performance. The advantage to the booster heater is that you can turn down your water heater thermostat, significantly reducing water heating costs. Resetting your water heater to 120°F (typically half-way between the "medium" and "low" settings) will provide adequate hot water for your household needs and reduce the risk of scalding. For more information, see Chapter 6.

Chapter 10

Laundry

Like dishwashers, most of the energy used to wash clothes is for heating the water. Water heating accounts for 70% to 90% of clothes washer energy use — even more than with dishwashers. The emergence of resource-efficient clothes washers on the American market has been an exciting development for consumers interested in energy savings and environmental quality. New resource-efficient washers provide excellent wash performance while cutting energy and water use by half or even more. These resource-efficient models also use higher spin speeds to extract more water from each load of clothes, saving significant dryer energy as a result. Unlike dishwashers, clothes washers can use cooler temperature water with perfectly adequate results for most clothes, further reducing energy use.

Buying a New Clothes Washer

Recent federal standards, effective January 2007, have improved the efficiency of all washers on the market by capturing a portion of the energy and water savings from resource-efficient designs. Still, it is usually not practical to replace your clothes washer before it fails on its own — typically after 10 to 12 years. When you are ready to buy a new machine, you may find a range of designs and special features to choose from, along with a range of costs to consider. As with other home appliances, it is always important to evaluate the long-term economics of purchasing a high-efficiency washer. A growing number of energy and water utilities around the country recognize the benefits of efficient clothes washers, and offer rebates to consumers who purchase qualifying machines. Call your energy and water utilities and ask if they provide rebates to offset the upfront cost for a high-efficiency washer.

For More Information:

To get the latest information on reducing the energy you use for laundry, go to www.aceee.org/consumerguide/laundry.htm. To identify the most efficient clothes washers, go to the ENERGY STAR website and search under "products."
www.energystar.gov ▪ 888-STAR-YES

New ENERGY STAR-qualified clothes washers must meet energy use and water consumption requirements — look for the ENERGY STAR logo as the first step in selecting your new washer. Once you have narrowed the field to include only ENERGY STAR models, it is important to realize that these models still vary considerably in total energy and water consumption. The table below illustrates the wide range of water and energy use for small, medium, and large capacity ENERGY STAR washers. To further evaluate the energy and water use of your preferred models, check out the detailed information available in the list of qualified clothes washers available at www.energystar.gov. You can also consult the EnergyGuide label and manufacturer literature.

TABLE 10.1 Range of Energy and Water Use for ENERGY STAR Washers

CLOTHES WASHER CAPACITY	ANNUAL ENERGY USE kWh/yr	$/yr	ANNUAL WATER USE Gallons/yr	$/yr*
Less than 2.0 cubic feet	Least 113	$10.75	2,766	$121.70
	Most 298	$28.30	5,749	$252.95
2.0 to 3.0 cubic feet	Least 126	$11.95	3,547	$156.05
	Most 257	$24.40	9,196	$404.60
More than 3.0 cubic feet	Least 120	$11.40	4,323	$190.20
	Most 316	$30.00	11,116	$489.10
Compare to Non-ENERGY STAR 3.2 cubic feet	463	$44.00	13,312	$585.73

* Assumes 9.5¢/kWh and 4.4¢ per gallon.

All new clothes washers must display EnergyGuide labels to help consumers compare annual energy use. The EnergyGuide label for clothes washers is based on estimated energy use for 8 loads a week — a total of 416 loads of laundry per year. But this value does not tell you the whole story for washers because of variations in tub size and other factors. Standard capacity washers range in size from 1.6 to more than 3.5 cubic feet. Conventional washers with smaller tubs may have better EnergyGuide ratings, but the smaller capacity may mean you have to run the machine more often, so it may actually cost more to operate.

> ## Did You Know?
>
> For each whole number drop in a washer's water factor — say, from 8.0 to 7.0 — you'll save about 1,000 gallons of water each year.

At present, many of the most resource-efficient clothes washers have a higher purchase cost than conventional washers; however, their substantial energy and water savings translate into big money savings and a quick return on your investment. Depending on your local energy and water rates and the amount of laundry you do each year, you may realize annual savings of $100 or more. If a resource-efficient washer costs $500 more to purchase than a conventional machine, your savings would be a tax-free return on your investment of 20%. Field studies have also shown that resource-efficient washers are gentler on clothes. Less dryer time also reduces wear-and-tear. With the average load of laundry valued up to $500, this can add up to substantial additional savings.

■ Front- vs. Top-Loading Clothes Washers

In general, horizontal-axis (usually front-loading) washers are much more efficient than conventional vertical-axis (top-loading) washers with agitators. However, several new top-loading designs achieve substantial energy and water savings compared to conventional top-loaders. To understand how these washers use so much less water and energy, consider the differences in washer design. In a conventional top-loader, the tub must be filled with water so that all the clothes are submerged. The agitator then swirls the water around, moving clothes against the agitator to clean them. In contrast, resource-efficient washers need less water because the tub never needs to be filled completely. In H-axis washers, the tub itself rotates, making the clothes tumble into the water.

Redesigned V-axis washers use sprayers to wet the clothes from above and a moving plate in the bottom of the tub to lift and bounce clothes through the wash water instead of an agitator. Very few agitator models qualify for ENERGY STAR.

Although the tubs in front-loading machines are often smaller than top-loading tubs, their capacity for clothes is often the same because the absence of agitators in the front-loading models allows their tumbling tubs to accommodate more clothes. Resource-efficient top-loaders that do not have agitators typically have a very large tub capacity as well. If you have a small family, or typically wash small loads, a smaller capacity front-loader may serve your needs better than a large washer.

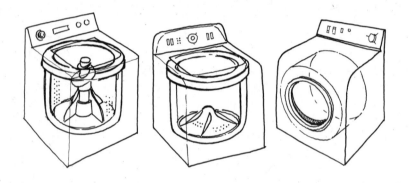

Horizontal-axis washing machines are far more energy efficient than conventional top-loaders. Some new top-loaders with improved designs are also energy efficient.

■ Wash and Rinse Cycle Options

Choose a clothes washer that offers plenty of choices for energy-conserving wash and rinse cycles. Wash and rinse temperatures have a dramatic impact on overall energy use — a hot water wash with warm rinse costs 5 to 10 times more than a cold wash and rinse. Cold wash cycles generally clean clothes perfectly well and are in fact recommended for many fabrics. Note: With oily stains, hot water may be required for satisfactory cleaning. You should experiment with the different cycle options and find one that meets your needs. Cold water rinses are just as effective as hot or warm rinses.

■ Water Level Controls

Most conventional clothes washers use approximately 40 gallons of water for a complete wash cycle. Large capacity resource-efficient models use less than 25 gallons per cycle; small and medium-sized models may even use less than 10. All front loaders and many of the higher-efficiency top-loaders feature advanced electronic controls to adjust the water level automatically according to the size of the load. If the models you are considering do not have these controls, choose a machine that lets you select lower water levels when you are doing smaller loads. For a given temperature cycle, energy use is almost directly proportional to hot water use. The lowest setting may use just half as much water as the highest. In general, you'll save energy by running one large load instead of two medium loads. Unfortunately, most manufacturers do not publish the actual water use of their machines in different settings, so it is difficult to compare one brand to another.

■ Faster Spin Speed

Faster spin speeds can result in better water extraction and thus reduce the energy required for drying. This is because mechanical water extraction by spinning is much more efficient than thermal extraction (heating clothes in a dryer). Front-loading washers and re-designed efficient top-loading machines generally spin clothes 2-3 times faster than conventional clothes washers, reducing the energy it takes to dry your clothes 5-15%. This also means you are saving time, whether you use a clothes dryer or hang your clothes out to dry.

Buying a New Dryer

Operation of dryers, both electric and gas-fired, is pretty straightforward: they work by heating and aerating clothes. Dryers are not regulated by the government so there is no requirement to display the EnergyGuide label on clothes dryers and no ENERGY STAR program for them. From an energy perspective, it makes little sense to replace a well-functioning dryer before the end of its useful life — typically 12 or 13 years. But even though the efficiency of dryers currently on the market does not vary widely, the features on a dryer and the way you use and maintain the dryer can have a big impact on energy use.

■ Gas vs. Electric

In terms of energy use, the performance of electric and gas dryers does not vary widely. Gas dryers may be less expensive to operate than electric models, but this of course depends on energy costs in your area. Electric ignition is required for all new gas dryers, eliminating the standing pilot lights found on older models as well as the associated energy waste and safety concerns. The advantage of using gas is that you are using the fuel directly on site, rather than allowing energy losses associated with generating and distributing electricity, which impacts the environment. One downside to gas is the potential introduction of combustion byproducts in your home if the unit is not vented properly.

■ Automatic Shut-off

Other than fuel type, the major energy consideration is whether the dryer uses termination controls to sense dryness and turn off automatically and, if so, the sensing mechanism used. Dryers used to be controlled simply by setting a timer. You guessed how long it would take to dry a particular load of laundry and set the timer for 60 minutes or 75 minutes, etc. If the clothes came out dry, chances are you continued to use that setting. However, some types of clothes dry much more quickly than others. You can save a significant amount of energy by buying a model that senses dryness and automatically shuts off. Most of the better quality dryers today include this feature.

> ### Did You Know?
>
> Clothes dryers use 2 to 4 times more energy than a new clothes washer, and almost twice as much electricity as a new refrigerator. High-efficiency washers save dryer energy use by removing more water in the spin cycle.

The best dryers have moisture sensors in the drum for sensing dryness, while others only infer dryness by sensing the temperature of the exhaust air. The lower-cost, thermostat-controlled models may overdry some types of clothes, but even these are much better than timed-dry machines. Compared with timed drying, you can save about 10% with a temperature-sensing control, and 15% with a moisture-sensing control. Overdrying can reduce the life of clothes, so getting the timing right saves money in your clothing budget as well.

■ Features That Reduce the Need for Ironing

While the energy used for ironing clothes is not technically part of the energy used for washing or drying clothes, it is closely related. If clothes are taken out of the dryer while still slightly damp and then hung up, they may not need ironing. Moisture-sensing automatic shut-off helps here — especially if the dryer includes a feature to adjust the level of dryness desired. Wrinkling can also be reduced with a cool-down (fluff) cycle and by a feature that tumbles the clothes periodically after the end of the cycle if the clothes are not removed right away. Look for these features when shopping for dryers.

Installing Clothes Washers and Dryers

Try to install your clothes washer as close to the water heater as reasonably possible and insulate the hot water pipes leading to it to minimize heat loss through the pipes. Also, if possible, locate the washer and dryer in a heated space. This is particularly important with dryers, which depend on heat to dry.

The most important part of the dryer installation is the exhaust system. In most cases — and always with gas dryers — the exhaust should vent to the outside, using as short and straight a section of smooth metal ducting as possible. Ducts that are unobstructed and minimize friction encourage better airflow, which improves efficiency and drying time. Take care when using flexible duct because if compressed or crushed behind the dryer, air flow will be restricted. If you do use flex duct, only use foil or aluminum types; plastic or vinyl will not hold up to the high temperatures of the dryer. Dryer vent hoods are available that seal very tightly when the dryer blower is not on. Standard dryer vent hoods have a simpler flapper that is not as effective. Ask your appliance salesperson about dryer vent systems and spend the extra $15 to $20 to buy a tight-sealing one.

Electric dryers may be vented inside the home during the winter months if the house air is dry and if the air vent is properly filtered. Compact electric dryers in apartments and condos are sometimes installed for indoor venting. If the electric dryer is vented inside, watch for moisture on the windows, which would indicate that you're introducing too much moisture into the house.

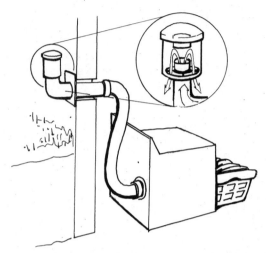

Put in a dryer vent hood that keeps out cold air.

Laundry Tips for Maximum Energy Savings

There are a number of easy ways to save energy with laundry, whether you're buying new appliances or not. Follow these suggestions whenever possible to keep energy use to a minimum. In most cases, practices that save energy also extend the life of your clothes.

■ Optimize Load Size

It is important not to underload or overload either your washer or dryer. Most people tend to underload their washers rather than overload — particularly with conventional top loaders — to make sure all the clothes are covered with water. Try to load your washer to its full capacity whenever possible without overloading. If you overload, clothes may not get clean and you may end up washing the load a second time.

If you are unsure about the size of a load, check your machine's load capacity in pounds, then use a household scale to weigh out a few loads of laundry to get a sense of how much laundry 10 or 18 or 20 pounds represents. Then use your eye to judge the volume of clothes for a load. Washing one large load will take less energy than washing two loads on a low or medium setting.

Also dry full loads when possible, but be careful not to overfill the dryer. Drying small loads wastes energy, while overloading causes wrinkling and uneven drying. Air should be able to circulate freely around the clothes as they tumble. If your washer and dryer are properly matched, a full washer load will be about the right size for the dryer as well.

■ Use Lower Temperature Settings

Use cold water for the wash cycle instead of warm or hot (except for greasy stains), and only use cold for rinses. Experiment with different laundry detergents to find one that works well with cooler water. By presoaking heavily soiled clothes, a cooler wash temperature may be fine. The temperature of the rinse water does not affect cleaning, so always set the washer on cold water rinse.

■ Use Energy-Saving Features

If your dryer has a setting for auto-dry, be sure to use it instead of the timer, to avoid wasting energy and overdrying, which can cause shrinkage, generate static electricity, and shorten the life of your clothes.

■ While You Dry

If you have the means of air-drying your laundry, this will save energy and wear and tear on your clothes! Enjoy the fresh air while hanging clothes on the line and use totally free solar energy for drying. If this isn't feasible, consider these quick tips:

Separate fabrics. Dry similar types of clothes together. Lightweight synthetics, for example, dry much more quickly than bath towels and natural fibers. Don't add wet items to a load that is already partially dried.

Dry two or more loads in a row. This takes advantage of the heat still in the dryer from the first load.

Try dryer balls. Increase air circulation and decrease drying time by purchasing dryer balls that bounce around with the clothes and fluff them as you dry. They can be reused indefinitely and are a chemical-free replacement for dryer sheets.

■ Proper Dryer Maintenance

Clean the dryer filter after each use. A clogged filter will restrict airflow and reduce dryer performance.

Check the outside dryer exhaust vent. If you have a conventional exhaust vent, make sure it is clean and that the flapper on the outside hood opens and closes freely. If the flapper stays open, cold air will blow into your house through the dryer and increase heating costs. Better yet, replace the outside dryer vent hood with one that seals tightly.

Chapter 11

Lighting

Lighting accounts for approximately 5–10% of total energy use in the average American home, costing the typical household between $75 and $200 per year in electricity. That's not a huge amount, but it is enough to justify doing something about — especially when the advantages of energy-efficient alternatives are considered. Making the switch to energy-efficient lighting is one of the quickest, easiest and least expensive ways to cut your home's energy use.

This chapter describes different types of lighting, when and how to use each to maximize energy savings and comfort, and how to find the right products. It also discusses the most important new developments in energy-efficient lighting. One note on terminology: The lighting industry uses the term lamp to refer to the actual source of light — what the public usually calls the light bulb. In this chapter, we use the terms lamp and bulb interchangeably.

For More Information:

To get the latest on reducing the energy you use for lighting, go to www.aceee.org/consumerguide/lighting.htm.

To identify the most efficient products, go to the ENERGY STAR website and search under "products" for lighting.
www.energystar.gov ■ (888) STAR-YES

Types of Lighting

■ Incandescent Lighting

Most lighting found in homes today is incandescent lighting. In an incandescent lamp, electric current heats up a metal filament in the light bulb, making it glow white-hot and give off light. The problem is that only 10% of that electricity is actually used to produce light — the rest ends up as heat. During the winter months, incandescent lighting is an expensive form of electric heat; during the summer months, it makes your air conditioner work harder than it has to.

We strongly recommend replacement of incandescent bulbs with high-efficiency alternative technologies like compact fluorescents. The energy, environmental, and economic benefits of CFLs are discussed in detail below.

■ Halogen Lighting

A halogen lamp is really a specialized type of incandescent lamp. A variety of halogen lamp types are available. Halogen is often used where high light quality or precise light focusing is required. Many halogen lamps feature a parabolic aluminized reflector (PAR) to improve light focus. Halogen lamps are slightly more energy-efficient than standard incandescent lamps, but not as efficient as CFLs. New halogen lamps designed as replacements for traditional incandescents are being introduced; however, they cannot match the significant energy savings of CFLs or the overall cost savings given the dramatic drop in CFL prices. Still, halogens are a good choice for accent lighting to highlight artwork or architectural features in your home.

■ Compact Fluorescent Lamps

The introduction of compact fluorescent lamps (CFLs) in the early 1980s revolutionized lighting; but only in the past few years have they gained mainstream popularity. Improved light quality, greater product variety, lower cost, and aggressive marketing and promotion among energy agencies, utilities, lawmakers, environmental groups, and even major retailers have really helped boost the market for CFLs. Still, some consumers remain skeptical. So, what is the deal with these light bulbs? Do they offer the same light quality? Are they safe for your home?

CFLs are getting so much attention for the simple reason that they use just one-quarter to one-third as much electricity and last up to ten times longer than incandescent bulbs, while providing the same level of light. This high performance means they waste much less electricity in the form of heat.

Although it is common to find high quality CFLs in most hardware stores, big box retail stores, and a growing number of grocery and drug stores, the selection from store to store varies widely. And, because this is a new product replacing a technology that has been familiar for more than 100 years, you may not even know what type you will like best. Because of this confusion, many consumers have been turned off by CFLs. It is important that you know about different kinds and share your experiences with others.

Compact fluorescent lights last on average ten times as long as incandescent lights.

What To Look For. The ENERGY STAR label will help identify high-quality, energy-efficient compact fluorescent lamps. To qualify for the label, lamps must comply with specifications on energy use, light output, lamp life, and other operating characteristics. A full list of ENERGY STAR lamps is available at www.energystar.gov.

The familiar wattages we use to purchase incandescent lamps — 40W, 60W, 75W, and 100W — are not useful when selecting CFLs. While most manufacturers label their CFLs with the equivalent incandescent wattage (ENERGY STAR lamps are required to display it), it is useful to know how the different wattages translate in terms of light output (or lumens). Table 11.1 will help you match CFL and incandescent lamps to get the same light level.

TABLE 11.1 Equivalent Incandescent and Compact Fluorescent Wattages

Incandescent Light Bulbs	Minimum Light Output	Compact Fluorescent Light Bulbs
WATTS	LUMENS	WATTS
40	450	9–13
60	800	13–15
75	1,100	18–25
100	1,600	23–30
150	2,600	30–52

You can find CFLs in styles to replace virtually any incandescent you currently use: traditional glass-encased bulbs, globe-shaped vanity bulbs, and flame-shaped bulbs with smaller screw bases for your candelabras and sconces. The majority of CFLs sold today are newer "mini" CFLs, which are the exact dimensions of an incandescent bulb so they will fit into most existing lamp fixtures. You can also buy three-way CFLs, dimmable CFLs, and bulbs adequate for use in low temperatures and moisture (but be sure to check

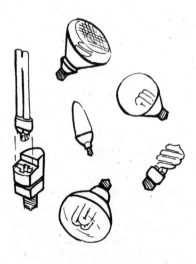

Compact fluorescents are available for many applications.

the label). Many of these specialty bulbs are only available in larger hardware stores and retail warehouses.

CFLs also come in a variety of color temperatures, meaning some give off "cooler" white light and some "warmer" white light. The shade of white light emitted by a lamp is called correlated color temperature — look for the value reported in Kelvin (K) on the lamp packaging. Lower Kelvin values (in the range of 2700K to 3000K) correspond to a warmer light comparable to a conventional incandescent that enhances yellow and red tones. Lamps with higher values (typically 3500K, 4100K, 5000K, and 6500K) give off a cooler white light which enhances blues and greens. These bulbs are often referred to as natural, bright white, or daylight lamps. The following table explains some terms you may encounter on an ENERGY STAR-rated CFL package. Try a few out with your family and see which you prefer. For instance, you may like a "soft white" for your living room and "cool white" for your bathroom. If the store where you are searching does not have a wide enough selection, suggest that they provide more offerings and try somewhere else.

TABLE 11.2 The CFL Light Color Spectrum

Color Term	Temperature (Kelvin)	Interpretation
"Warm white"	2700 K	Just whiter than a candle flame (yellowest)
"Soft white"	2800 K	Like a regular 60W incandescent bulb
"Bright white" or "Medium white"	3000 K	Neutral white, as in new offices
"Cool white"	4100 K	Clean, bright, white light.
"Daylight"	5300 K	Approximates sunlight on a bright day (bluest)

If you are shopping for specialty CFLs, be sure to use them according to manufacturer instructions. Never use compact fluorescent lamps in circuits that have dimmers unless the lights are specifically designed for that use. For recessed downlights, spotlights, and track lights, some compact fluorescent lamps are too wide at the base to fit into the can or cone. Again, finding a product that fits may take some trial and error.

Did You Know?

Energy-efficient light fixtures are available in a wide range of styles for use with CFLs, linear fluorescents, and LEDs. Fixtures designed specifically for use with energy-efficient bulbs will distribute light most evenly.

Environmental Benefits. A single 20-watt compact fluorescent lamp, used in place of a 75-watt incandescent light bulb, will save about 450 kWh over its lifetime. That single replacement will eliminate emissions of 700 pounds of carbon dioxide, 4 pounds of sulfur dioxide, and 6 mg of mercury on average. If your electricity is generated in a coal-fired power plant — half of all electricity produced in the U.S. — the emissions reductions will be roughly double.

Economics. Because CFLs save electricity and last longer, they are a profitable investment. To figure out your return, you have to look at both the purchase and operating costs to arrive at total life-cycle costs. It may surprise you to learn that incandescent bulbs cost a lot more to operate than they cost to buy. When you spend 50¢ for a 100-watt incandescent light bulb, for example, you're committing yourself to spending more than $7.00 for electricity (at 9.5¢/kWh) over that bulb's expected 750-hour life. In regions with high electricity prices, you could spend up to twice as much.

With a high-quality compact fluorescent lamp, you might spend anywhere from $2 to $5 to buy the lamp, but you'll save money in the long run. Most compact fluorescent lamps last about 6,000–10,000 hours. ENERGY STAR-qualified CFLs list their rated lifetime on the packaging. The longer a particular light fixture is used each day, the faster a compact fluorescent replacement will pay for itself. See for yourself in the following table. This assumes a $4 price per CFL, but you may find much lower prices per unit. Even with just one switch, you come out ahead in the first year. Just imagine how much money you would save if you switched all your bulbs to CFLs!

TABLE 11.3 Dollar Savings Achieved by Switching from Incandescent to Compact Fluorescent

Replace 75-watt incandescent with 20-watt CFL	Savings after 1st year	Savings after 2nd year	Savings after 3rd year	Savings after 5th year	Savings after 10th year
Light on 2 hrs/day	$ 0.31	$ 4.63	$ 8.94	$ 17.57	$ 39.14
Light on 4 hrs/day	$ 4.63	$ 13.26	$ 21.89	$ 39.14	$ 78.29
Light on 8 hrs/day	$13.26	$30.51	$43.77	$78.29	$ 156.07
Light on 12 hrs/day	$ 21.89	$43.77	$ 69.66	$ 117.43	$234.36

ASSUMPTIONS:

	Incandescent	Fluorescent
Lamp output (lumens)	1200	1200
Lamp life (hours)	750	8,000
Lamp cost ($)	$0.50	$4.00
Electricity cost	9.5 ¢/kWh	

CFLs, Mercury, and Safety. Compact fluorescent lamps contain a small amount of mercury vapor that is needed to produce light (about 5 mg). The amount is small enough that it presents very little danger during household use — even if the lamp breaks. However, lamps should be disposed of properly to minimize the spread of mercury in your home or in landfills. The ENERGY STAR Lighting Program is working with manufacturers, retailers, and local communities to establish recycling programs for spent compact fluorescent lamps.

For More Information:

For tips on how to recycle your CFLs and what to do if one breaks, try these resources.

www.lamprecycle.org   www.energystar.gov
1-877-EARTH911

Common CFL Myths

- **The light is green.**
  CFLs are not like incandescent bulbs; they do not emit full-spectrum white light. Instead, the color must be manufactured, which can yield different results from product to product. ENERGY STAR-qualified warm white and soft white CFLs closely approximate the light of a typical incandescent bulb.

- **Don't they flicker?**
  Today's CFLs use electronic ballasts that eliminate the flicker common to older CFLs and linear fluorescent systems using magnetic ballasts.

- **CFLs aren't bright enough.**
  An incandescent bulb of a higher wattage (say 100W) has a slightly higher, "bluer" apparent color temperature. If you think your CFL is too dim no matter what wattage you try, try a higher color temperature instead.

- **It is best to leave CFLs on all the time.**
  Turning CFLs on and off does not significantly reduce lamp life.

Integral vs. Modular Compact Fluorescent Lamps. All fluorescent lamps need ballasts to operate. The ballast is a device that alters the electric current flowing through the tube, which activates the gas inside, causing the tube to emit light. For this reason, compact fluorescent lamps are more complex to manufacture than standard incandescent light bulbs, and until recently they were a lot larger. Most compact fluorescent lamps today have electronic ballasts that are integral with the lamp. That is, the ballast and lamp are combined in a single unit, which screws into a standard light bulb socket.

Modular compact fluorescent lamps have separate ballasts and lamps. Some modular ballasts, or adapters, screw into standard light bulb sockets. Others are hard-wired into light fixtures — i.e., the ballast is built into the fixture. The advantage of modular compact fluorescent lamps is that you don't have to replace the ballast when the fluorescent tube fails. Compact fluorescent lamps generally last about 10,000 hours, while ballasts are expected to last 50,000 hours or longer.

■ Linear Fluorescent Lighting

When you think of standard tube fluorescent lighting, you may get an image of the buzzing, flickering bluish-white lights in supermarkets or older offices that make colors look washed out and give some employees headaches. That is hardly the kind of light you want in your house. Well, times have changed. Linear (tube) fluorescent lighting has improved dramatically over the past 20 years. The best lamps with electronic ballasts are a far cry from what most of us think of as fluorescent lighting. They now make sense in places other than your garage or basement workshop. In fact, they can provide very satisfactory (and energy-efficient) recessed lighting around the perimeter of a living room, or overhead lighting in kitchens and bathrooms. You can now even buy dimming fluorescent fixtures to vary ambient light levels precisely for excellent mood lighting. For use in living areas, ask for fixtures with electronic ballasts that save as much as 35% on energy use. These cost more than standard magnetic ballasts that are still common in residential fixtures, but they eliminate any noticeable hum or flicker, making them far more pleasant in your home environment.

When shopping for linear fluorescent lighting fixtures and lamps, it helps to know what you want. Even at retail lighting stores, salespeople may not be familiar with some of the advanced products on the market. If you can't find what you want, go to a commercial lighting supplier. Commercial-grade fluorescents offer higher quality and greater selection.

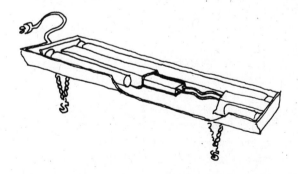

**Linear fluorescent fixture showing cutaway lamps and ballast.
You can buy quiet, flicker-free electronic ballasts for these fixtures.**

When selecting linear fluorescent lighting, look for products with a high color rendition index (CRI), that is, 75–90 CRI, or the higher the better. CRI is a measure of the ability of the light to illuminate colors accurately. Also, look for high efficiency (or efficacy, as it is called in the lighting industry). Lighting efficacy is measured in lumens (light output) per watt (electricity use). The best fluorescent lamps use special coatings ("trichromatic phosphors") to achieve both high CRI ratings and high efficacy. The most energy-conserving fluorescent lamps are thinner in diameter ("T8" and "T5") and may require different fixtures and ballasts than what you are used to ("T12").

■ HID Lighting

High-intensity discharge (HID) lighting is what you typically see along streets and in parking lots. HID lighting has advanced almost as quickly as fluorescent lighting in recent years. There are three types commonly used: mercury vapor, high-pressure sodium, and metal halide. Like fluorescent lamps, they require ballasts to operate, and most HID lamps take several minutes to warm up. The primary household use of HID lighting is outdoors: to light up the driveway, swimming pool, tennis court, etc.

While mercury vapor lamps are still used for outdoor lighting, they are quickly becoming obsolete because of the higher efficacy of high-pressure sodium and metal halide. High-pressure sodium lamps are available with efficacies as high as 140 lumens per watt, though the light is somewhat yellowish. Metal halide lights produce a whiter light, closer to incandescent in quality, with relatively high efficacy.

■ LEDs (Light Emitting Diodes)

Light emitting diode (LED) technology is evolving rapidly as a new, high-efficiency option for a variety of residential lighting applications. An LED is a pea-sized device that uses semi-conducting metal alloys to convert electricity into light. These diodes can be arranged in matrices or clusters offering great design flexibility. Over the past few decades, the development of first blue and now white LEDs has extended their use from simple indicator lights to a wider range of display and task light applications, from traffic signals and automobile brake lights, to TV screens. White LED products on the market for home use today include desk lamps, under-cabinet lights, flashlights, outdoor pathway lights, and decorative string lights. Most of these products have reached the market only within the past 5 years.

Light emitting diodes (LEDs) use semi-conductor technology. They are more durable and last longer than other lighting technologies.

So should you rush out and buy an LED lamp? Not necessarily. Because white LEDs are still in their infancy for general applications, performance varies from product to product. The best white LEDs are similar in efficacy to compact fluorescent lamps and improving year by year; as of 2007 the majority of available consumer products are only a little better than incandescent lamps, but cost a lot more. Even though they use little energy, light output of some products can deteriorate rapidly. To understand whether an LED product is appropriate for you, consider the following:

LEDs are currently sold for several applications and may be a good replacement for task-oriented fixtures.

- **Efficacy.** Bright LED lamps should be able to maintain at least 30–35 lumens per watt (lpw). Incandescent lamps provide about 15 lpw and compact fluorescents offer 50 lpw. Be sure to check and verify performance with the manufacturer.

- **Light Focus.** LEDs are very good at directing light in a specific direction. So even if they are not able to provide the same measured total light output per watt as another technology, LEDs may offer superior illumination for a particular task. In contrast, incandescent and fluorescent lamps emit light in all directions, which means a lot of light is absorbed inside the fixture, as in a recessed ceiling fixture.

- **Durability.** LEDs are a great option for flashlights, head lamps, and outdoor walkway lamps. They not only use very little power and last longer, they are also shatterproof and shock resistant, and their performance is not degraded by low temperatures or moisture.

- **Decoration.** During the holidays, most utilities experience a significant surge of electricity usage from the lighting of Christmas trees, yards, and other ornamentation. That means more pollution in the air and more of your money to the utility. LED string lights are now widely available in white and a range of colors including multi-color sets, providing a more durable and low-energy alternative. They may cost more up front, but one utility in California estimates that using LED lamps will cost you just $0.45 over the course of a holiday season, compared to $5.00 for mini incandescent string lights and $75 for large string lights!

- **Color.** Be aware that the best LEDs on the market provide a cooler, bluish light than incandescent and fluorescent lamps. Typically, high performance is more difficult to achieve with warmer whites.

The government is working hard to develop a standard for identifying and certifying quality LED products, which will provide consumers with better performance and reliability. Although not yet available, keep your eye out for products that carry the ENERGY STAR label. Meanwhile, LEDs are sure to continue to improve exponentially, and savvy consumers can expect to see white LEDs in a growing number of applications at efficacies comparable to fluorescent technology within the next 5 to 10 years.

Light Fixtures

■ Dedicated Compact Fluorescent Fixtures

Efficient lighting goes beyond selecting the right light bulb for your
existing fixtures. ENERGY STAR
endorses a wide range of indoor and
outdoor fixtures that are designed for
use with CFLs and linear fluorescent
lamps. These fixtures are available in
most home improvement centers and
lighting showrooms. Many of the
indoor fixtures incorporate dimmers
or two-way switches. Outdoor fixtures automatically shut off during the
day, or come equipped with motion
sensors.

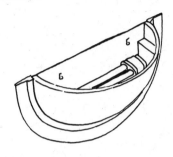

**Hard-wired fixture for
compact flourescent lamp.**

■ Ceiling Fan Fixtures

ENERGY STAR also endorses high-efficiency ceiling fan/light combination units and ceiling fan light kits. Because the lighting component
represents a greater energy savings potential than the fan, be sure that
your ENERGY STAR ceiling fan is also equipped with a qualified light kit.
An efficient light kit can reduce ceiling fan energy consumption by 60%
to 80%.

■ Torchiere Lamps

It is important to distinguish halogen torchiere floor lamps from the other
halogen lighting discussed above. These fixtures, which became
popular in the 1990s for providing pleasing, indirect light at a low upfront cost, are actually quite inefficient and costly, since they consume
300–600 watts of electricity yet direct the light to the ceiling. Even
though a smooth, white ceiling can reflect some light, bathing a ceiling

For More Information:

Visit ENERGY STAR for a full listing of qualified residential
light fixtures, ceiling fans and torchieres, as well as
information on manufacturers and retailers.
www.energystar.gov ■ (888) STAR-YES

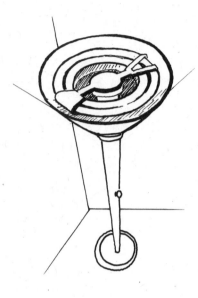

Replace your old halogen torchiere with an ENERGY STAR fluorescent torchiere.

with light from a halogen torchiere wastes the primary benefits of halogen lamps. Halogen torchieres are particularly poor choices in rooms with non-white or textured ceilings. In addition, these light fixtures pose a fire hazard due to the extremely hot temperatures produced by their high-wattage bulbs.

Fortunately, as of January 2006, efficiency standards ban the manufacture of lamps that consume more than 190 watts, thereby eliminating harmful halogen torchieres. Although they are being phased out in retail stores, they may still be available for purchase and should be avoided. If you currently have a halogen torchiere in your home, replace it with a torchiere designed for use with fluorescent lamps. Several companies make attractive and energy-efficient fluorescent torchieres that feature full-dimming or three-stage dimming (three light levels) and are much safer than halogen floor lamps, while using only a fraction of the electricity. Most of these products carry the ENERGY STAR label for easy identification. Check with your utility to see if it will buy back your halogen torchiere or offers an incentive for a new ENERGY STAR-qualified lamp.

Did You Know?

Halogen torchieres can reach temperatures of 950°F to 1200°F creating a real fire hazard and more work for your air conditioner. Efficient fluorescent torchieres offer the same high-quality, adjustable lighting along with substantial energy savings without the safety concerns.

■ Outdoor Lighting

The lights you use to illuminate your walkway, driveway, patio and front or back doors can have a significant impact on your energy bill as well as on others in your community, especially if you tend to leave the lights on through the night or (horrors) during the day. The main difference between saving energy with indoor and outdoor lighting is that more efficient outdoor fixtures can benefit both your pocketbook and the safety of your home and community. Poorly designed outdoor lighting, which is brighter than necessary, illuminates too broad an area, or directs light skyward or next door, can create glare and deep shadows, irritate neighbors and even damage surrounding wildlife! A growing body of research indicates that artificial night lighting that disrupts natural exposure to "dark hours" can be detrimental to nocturnal ecosystems and even to human health.

In order to have energy-efficient and visually effective outdoor lighting, first determine if there is a true need for light, and then consider the purpose and placement of the light. Where needed, fixtures should shield the light source from view to eliminate glare, they should be directed properly to the ground, and they should employ efficient light sources at appropriate wattage levels. Suitable technologies include compact fluorescents that can operate at cold temperatures, LEDs, ceramic metal halides and high-pressure sodium lamps. Light your property to avoid areas of deep shadows because they are difficult to see into and provide criminal-friendly hiding places. Additionally, security and landscape lighting usually do not need to be on all night. Motion controls, infrared sensors, dimmers, and time controls are cost-effective tools to help eliminate unnecessary attention toward your property, particularly in the hours after midnight.

For More Information:

To get an idea of the types of fixtures to use, the International Dark Sky Association certifies dark-sky friendly fixture products and lists manufacturers that are also a good choice for the energy-conscious.
www.darksky.org/programs/fixture-seal-of-approval.php

■ Solar-Powered Walkway and Patio Lights

Another novel idea in lighting is the use of solar energy to power outdoor lights. During the daytime, a photovoltaic (PV) panel generates electricity, which is stored in a battery. At night, the stored electricity is used to power the light. Some models are turned on manually, while others are turned on automatically by light-sensing controls or activated by motion-sensing devices. Most of these walkway or security lights require no wiring or installation other than pushing a stake into the ground or screwing the fixture to a garage wall.

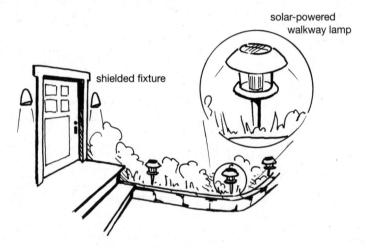

solar-powered
walkway lamp

shielded fixture

Outdoor lighting should be shielded and take advantage of solar power when possible.

Most of the widely marketed solar walkway lights do not put out a whole lot of light — don't expect to read the newspaper under a $30 walkway light — but they are very useful for lighting the path to your door so your guests can find their way. Larger solar lights are available that do provide a lot of light, but these can be quite expensive.

Solar-powered outdoor lights can be found in many hardware and department stores or purchased through catalog retailers of alternative energy and stand-alone power equipment. You can install them yourself in a few minutes without having to bury wires or hire an electrician.

Improving Your Lighting Efficiency

■ Lighting Controls

In addition to saving energy by using more energy-efficient lamps, you can also save by leaving the lights on for a shorter period of time or dimming them. The simplest lighting control strategy, of course, is to turn lights off when you leave a room. Even if you are leaving for just a few minutes, it saves energy to turn the lights off. This applies to both incandescent and fluorescent lights. The number and variety of controls available is growing as the technology develops along with consumer interest in home automation. Check to make sure inexpensive controls are compatible with your electronic ballasts.

With closets and cabinets, you can buy switches that turn the lights on when the door is opened and turn them off when the door is closed, thereby saving energy — as long as you remember to shut the door.

If you're the type of person who forgets to turn off lights when you leave a room, consider installing occupancy sensors that automatically turn lights off once a room is vacant. Most of these work by sensing heat or motion. While used primarily in commercial buildings, they can save energy in the home as well.

Light-sensing controls are increasingly being used to control outdoor lights. If you currently turn outdoor lights on manually at night, these controls are a great convenience and they can save even more energy

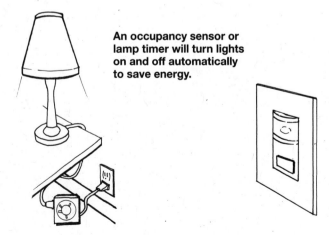

An occupancy sensor or lamp timer will turn lights on and off automatically to save energy.

if connected to a timer to turn the lights off sometime late at night. Motion-sensing controls for outdoor security lighting save energy and discourage potential intruders.

Another control strategy is to reduce light level and energy use with dimmers. Linear fluorescent lamps are dimmable only if they have dimming ballasts. Dimmable compact fluorescents have been introduced — check the packaging to determine whether a compact fluorescent will work with a dimmer. Dimming will reduce the energy consumed by fluorescent lamps without affecting efficiency. This is not so for incandescent lamps. Because of the way incandescent lights work, dimming to half the light output may still require three-quarters of the power, not half the power.

■ Use Daylight

Nothing's nicer than natural light, and in terms of energy use, nothing's more efficient. A single skylight or properly positioned window can provide as much light as dozens of light bulbs during the daylight hours. To benefit more from natural lighting you may need to rearrange your rooms somewhat — putting your favorite reading chair over by the south window, for example. Or you may want to go to more effort and

Natural daylighting is the least expensive light source of all.

install one or more skylights. To help get that light deeper into the room, you can paint your walls a light color and use reflective louvers or venetian blinds.

Another option is to install tubular skylights to bring daylight into rooms where traditional skylights aren't feasible. Tubular skylights transmit daylight through a cylindrical tube from a small dome on the roof through the attic to a ceiling-mounted diffuser. The tube is lined with a highly reflective surface that channels daylight to the interior. The diffuser looks much like a conventional ceiling-mounted light fixture.

As you plan for natural lighting, don't forget that too much glass area on the east or west walls or on a south-facing roof can increase your air conditioning requirements. The best design balances passive solar heating, daylighting, and cooling considerations.

■ Match Your Light to the Task

You can save a lot of energy by concentrating light just where it's needed and reducing background or ambient light levels. This strategy — called task lighting — is widely used in office buildings, but it makes just as much sense at home. Install track or recessed lights to illuminate your desk or the kitchen table where you do the crossword puzzle, and keep the ceiling lights off. Small compact fluorescent lamps can be used in many locations. New LED strip lights can focus additional illumination on countertops and other work surfaces.

Use task lighting where you need the light most and reduce ambient light levels.

Chapter 12

Home Electronics

Reading your monthly electric bill, you would know if you stopped using your air conditioner since the previous month, but you may not notice if you stopped watching television, because electronic appliances don't use as much energy by themselves. Their energy use often goes unnoticed. But as it turns out, an estimated 10% to 15% of all electricity used in American homes can be attributed to the buzz of electronic devices, not because they use a lot of energy, but because of their sheer numbers. As electronics use continues to proliferate, and as new products with higher energy demands hit the market, the overall portion of the household energy budget devoted to electronics is expected to grow.

For More Information:

To get the latest information on reducing the energy consumed by your home electronics, go to www.aceee.org/consumerguide/electronics.html.

To identify the most efficient products, go to the ENERGY STAR website and search under products for "home electronics" and "office equipment."

www.energystar.gov ▪ (888) STAR-YES

The vast majority of home electronics energy use — up to 90% by some estimates — is consumed by home entertainment systems and home office equipment. The remaining 10% consists of many small energy users, including portable devices with battery chargers. Although each of these products uses a relatively small amount of electricity on an individual basis, they continue to proliferate rapidly and represent an opportunity to keep overall electronics energy use in check.

Thinking About Power Modes

To minimize the energy used by home electronics, it is helpful first to understand their operating modes and how power consumption varies by mode. This will make it easier to understand the opportunities for saving energy in the many different electronics products you most likely have in your home.

For starters, it is helpful to understand the true meaning of "on" and "off" as applied to electronics. It's rarely that simple! Unlike a light switch that turns a lamp or fixture on or off, many electronics products operate in two, three, or even four modes, and even continue to draw power when apparently turned off. Commonly identified modes and their definitions are shown in Table 12.1. Home electronics use power in standby and off modes to support features such as instant-on, remote control, channel memory, and LED clock displays. Common electronics components, like power supplies and battery chargers, contribute to standby and off mode power consumption and also impact a product's in-use energy consumption.

Active mode power consumption varies widely by product type, from less than 10 watts for some computer peripherals and stereo equipment to more than 200 watts for plasma screen TVs. Recent studies estimate that active mode power accounts for approximately 60% of electronics power consumption. Low power modes (including active standby, passive standby, and off mode) typically range from less than 1 watt to around 10 watts although some products can use as much as 50 watts in one or more low power mode! The average U.S. household consumes 50 watts of standby and off-mode power constantly, amounting to about 440 kWh per year. Nationwide, this electricity costs consumers over $3 billion per year.

TABLE 12.1 Common Electronics Operating Modes

Mode	Definition	Examples
Active (In-Use)	Appliance is performing its primary function.	TV displays picture and/or sound. VCR records or plays back tape. Printer prints document.
Active standby	Appliance ready for use, but not performing primary function. Appears on to consumer.	DVD player on but not playing. Cordless appliance charging.
Passive standby	Appliance is off/standby. Appears off to consumer, but can be activated by remote control OR is performing peripheral function.	Microwave not in use, but clock is on. CD player off, but can be turned on with remote control.
Off	Applicance is turned off and no function is being performed. Consumer cannot activate with remote control.	Computer speakers are off, but plugged in. TV is not functioning and cannot be turned on with remote.

Power Supplies

Electronics products run on low-voltage direct current (DC) and therefore require power supplies to transform the 120-volt alternating current (AC) supplied at the power outlet. Some larger products, like TVs, stereos and set-top boxes, incorporate the power supply into the body of the product. Others use external power supplies, the familiar "wall packs" that increasingly compete for space in our outlets and power strips. These power supplies consume electricity whether the product is on or off — and even if the product is disconnected! The fact that these power supplies are drawing power can be demonstrated simply by grasping a wall pack that has been plugged in for a while: it will be warm to the touch. That warmth is simply electrical energy wasted as heat.

Did You Know?

The 2.5 billion power supplies in use throughout the U.S. consume around 2% of the country's electricity production and emit almost 50 million tons of carbon dioxide each year.

A number of manufacturers now offer high-efficiency power supplies (typically "switch-mode" power supplies) and a growing number of products are sold with these improved devices. The best of these devices boast efficiency levels of more than 90%, whereas the worst performers are only 20-40% efficient (meaning they waste more than half of the electricity that passes through them!). As a bonus, high-efficiency power supplies are much smaller and lighter than the wall-pack power supplies they replace, saving room under your desk and in your briefcase. ENERGY STAR-qualified power supplies are now available and are being sold with a growing number of electronics products.

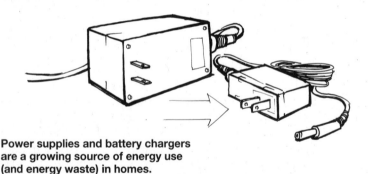

Power supplies and battery chargers are a growing source of energy use (and energy waste) in homes.

Home Entertainment Equipment

Considering that more than half of all American households have two or more TVs (and over five million households have four or more TVs), it is no surprise that home entertainment systems are a major household energy use. Table 12.2 lists the power consumption of common electronics products across their operating modes, as well as annual electricity use. Use the table to estimate how much energy is consumed by these electronics in your home.

TABLE 12.2 Energy Usage of Common Home Entertainment Appliances

Product	Passive Standby and/or Off	Active Standby	Active Power (in use)	Average Annual Energy Use
	WATTS			kWh
Televions				
Rear Projection	3	-	160	447
Plasma Screen TV (<40")	1	-	246	441
CRT Screen TV (<40")	3	-	73	123
LCD Screen TV (<40")	2	-	70	77
TV Peripherals				
DVR or TiVo	37	37	37	363
Digital Cable Box	26	26	26	239
Satellite Receiver	12	11	16	124
Analog Cable Box	10	10	10	89
VCR	2	7	13	34
Video Game Console	1	-	24	16
DVD player	1	5	11	13
Audio Equipment				
Audio Receiver	3	-	50	143
Audio Minisystem	6	-	14	58
Portable Stereo	2	6	8	18
Radio	2	2	12	18
Audio Amplifier	0	-	40	13
CD Player	1	-	7	12

Note: Average annual energy use assumes a typical number of hours spent in each mode.
Some appliances, like set top boxes, only have one or two modes, or the modes are indistinguishable.
Source: ECOS Consulting, 2006: *Final Field Research Report for the California*
    Energy Commission

Until recently, most TVs used the same technology, and power consumption did not vary greatly from product to product. Rear projection models were the exception, using 3 to 4 times more energy than traditional CRT (picture tube) models. Times are changing rapidly and many new TV technologies — mostly digital, flat-screen models — are now available and prices for these technologies are dropping. When you set out to purchase a new TV, it is useful to know what if any energy penalty that slick new set comes with. Two of the leading new tech-nologies — LCD and plasma — fall at opposite ends of the energy use spectrum. Pick an LCD model and you could cut your energy costs compared to your old TV, pick plasma and your bill could more than triple.

The expanding universe of set-top boxes is also contributing to growing home entertainment energy use. The biggest users are new digital products including digital video recorders (such as TiVo) and cable and satellite boxes that use two-way communication to send information back and forth between your TV and your service provider. DVDs, VCRs, and video game consoles contribute a small fraction to total entertainment energy use.

For close to a decade, manufacturers have worked with the U.S. EPA to develop home electronics with low standby losses. Look for the ENERGY STAR label when purchasing new home entertainment equipment. Simply through improved circuitry and better power supplies, ENERGY STAR products will use much less electricity than conventional electronics — in most cases only 1 watt or less. As standby power is ratcheted down, new technologies with higher power demands are introduced, increasing the average household use of entertainment equipment; thus, the contribution of active mode power consumption to overall energy use is growing for many products. The ENERGY STAR program is now working to develop new specifications for TVs that incorporate active mode power requirements.

The Coming Conversion to Digital Television

As of February 18, 2009, the U.S. will shift to digital-only TV broadcasts. Consumers who rely on over-the-air broadcasts (those who do not subscribe to cable or satellite services) will need a digital TV converter box (DTA) to view programming on their analog TVs. These boxes should become widely available on the market by mid- to late-2008. If you have purchased a digital TV, you will not need a DTA for that TV set.

Simple DTAs are expected to cost around $50. The National Telecommunications and Information Administration will offer consumers coupons to offset the cost of DTAs. Each household is eligible for two $40 coupons. DTAs eligible for the coupon program must meet energy efficiency specifications including a maximum standby power level and automatic power down after 4 hours of inactivity. Additional ENERGY STAR requirements set maximum active power levels for DTAs. To minimize your energy use, look for ENERGY STAR-labeled DTAs if you need to purchase one for your home.

For more information on the digital TV transition, go to www.dtv.gov/consumercorner.html.

Computers and Home Office Equipment

More and more people are working at home or using computers in their leisure time. As home office use increases, so does energy use by such equipment as computers, printers, copiers, fax machines, and other computer peripherals. Some of this equipment (especially color monitors and laser printers) consumes a great deal of energy. Table 12.3 summarizes typical power use by operating mode for office equipment and average annual electricity use. Note the big difference in energy use between laptop and desktop computers as well as the contribution of common computer peripherals, all things to keep in mind as you consider upgrading your equipment.

Look for the ENERGY STAR label on any new computer equipment you purchase. The ENERGY STAR label identifies efficient PCs, printers, faxes, and copiers. Current specifications set maximum power levels for sleep mode power consumption (and, in the case of monitors,

Small amounts of standby power can add up in the typical home office.

active mode power) as well as requirements for power management features. Remember, power management features for your computer and monitor must be enabled to save energy.

Fortunately, any increase in household energy use when you work at home is usually made up for by reduced energy use for transportation. In fact, the trend toward more home offices could have a significant impact on energy use for commuting.

TABLE 12.3 Energy Usage of Common Home Office Appliances

Product	Standby Power Use (off)	Low Power Use (on but inactive)	Active Power Use (on and performing)	Average Annual Energy Use
		WATTS		kWh
Computers				
Desktop Computer	4	17	68	255
Laptop Computer	1	3	22	83
Computer Monitor (CRT)	2	3	70	82
Computer Monitor (LCD)	1	2	27	70
Printing and Imaging				
Multi-function Device	6	9	15	55
Fax	4	4	4	26
Inkjet Printer	2	3	9	15
Laser Printer	1	10	39	15
Copier	1	-	18	11
Flatbed scanner	6	-	12	9
Small Computer Peripherals				
Modem	5	-	6	50
Wireless Router	2	-	6	48
Computer Speakers	2	-	7	20
USB Hub	1	-	3	18

Note: Average Annual Energy Use assumes a typical number of hours spent in each mode. Some appliances only have one or two modes, or the modes are indistinguishable.

Source: ECOS Consulting, 2006: Final Field Research Report for the California Energy Commission

Portable Electronics

Many rechargeable products are sold with simple battery chargers and power supplies that continue to draw power even after the product is fully recharged. An easy way to cut this power use is to unplug rechargeable appliances and devices when the battery is recharged or the product is not in use. Also, consider whether it is necessary to keep continuously charging the batteries for all of the rechargeable products in your home.

TABLE 12.4 Energy Usage of Common Rechargeable Appliances

Product	Standby Power Use (off but plugged in)	Low Power Use (on but inactive)	Active Power Use (on or recharging)	Average Annual Energy Use
	WATTS			kWh
Hand-Held Vacuum	3	-	3	29
Cordless Phone	2	3	5	26
Electric Toothbrush	2	-	4	14
Shaver	1	-	1	11
MP3 Player	1	-	1	6
Cell Phone	0	1	3	3
Digital Camera	0	-	2	3

Note: Average Annual Energy Use assumes a typical number of hours spent in each mode.
Some appliances only have one or two modes, or the modes are indistinguishable.
Source: ECOS Consulting, 2006: *Final Field Research Report for the California Energy Commission*

Reducing Home Electronics Energy Use

In addition to looking for the ENERGY STAR label when purchasing new electronics products, there are several steps you can take to minimize the energy used by the electronics in your home now.

Unplug It. The simplest and most obvious way to eliminate power losses is to unplug products when not in use. For example, unplug chargers for your portable devices when you take the products off.

Use a Power Strip. Plug home electronics and office equipment into a single power strip with an on/off switch. This will allow you to turn off all power to the devices in one easy step. But remember to keep your power strip in an easy-to-reach location! Once the power strip is turned off, no power will be delivered to the outlets, thereby eliminating power wasted by power supplies. One caveat: home entertainment equipment such as TVs, cable and satellite boxes, and DVRs will need to be reprogrammed or given time to reboot and download information when turned back on. You may want to plug these devices into a separate strip and only turn them off when you plan to be away for more than a few days.

Use a Power Meter. Inexpensive power meters are now available that can accurately gauge power consumption even at very low power levels. You plug the device in between the appliance and the wall socket, and watch the electricity use change as the appliance goes in and out of power modes. In addition to giving instant readings of power use, several of these devices will record energy consumption over the course of an hour, day, week, or even a year, and hook into your computer to show you a graph of the trends. Use a power meter to find your leading sources of energy consumption. This will help you to prioritize which products to unplug or to replace. Two models to look for are the Kill A Watt™ and the Watts Up? Pro Power Meter.

For an even more sophisticated, big-picture look at your home's real-time electricity use, you might also consider purchasing a power use monitor. These devices are programmed to read information from your electric meter and communicate the real-time changes in use through an easy-to-read screen. The best monitors are wireless and portable. When your clothes dryer turns on, you'll see the degree to which your electricity use spikes. When nothing is operating, you'll see what the background buzz of electric use is in your house, and try to track down the top appliances to be unplugged. Plus, power meters and real-time monitors can be a way to get your family involved and interested in saving energy. Some good monitors to look for are The Energy Detective (TED), the Power Cost Monitor, and the Cent-A-Meter.

For More Information:

A good selection of power meters and real-time monitors is available here.
www.powermeterstore.com ■ (877) 766-5412

Chapter 13

Other Energy Uses in the Home

When we think about saving energy in the home, we generally focus on the obvious energy uses: heating, cooling, water heating, refrigerators, and so on — uses and products that the bulk of this book addresses. In many homes, though, there are miscellaneous uses of energy large and small that are for the most part overlooked. We've already discussed some of these in Chapter 12. Some of the larger energy users are waterbed heaters, well-water pumps, pool filtering systems, engine block heaters, hot tubs, or even aquariums for tropical fish. While the energy use of each of these products may be relatively small on a national level, it can be quite large in individual houses. In fact, it is not too uncommon to find that one or more of these miscellaneous products can account for more energy use than your refrigerator, water heater, or even heating system.

Other common products, such as small kitchen appliances, recharge-able tools, and hard-wired loads (doorbells, security systems, garage door openers), use relatively small amounts of energy on an individual basis but consume a lot of electricity overall, due to their sheer numbers. And when these miscellaneous products are looked at collectively on a national level, their total energy consumption is significant — from 5% to 10% of residential electricity consumption — and growing.

Let's take a look at these miscellaneous energy uses — both how they can affect energy use in an individual home and what the nationwide effect is. Table 13.1 lists the typical energy consumption of a subset of these household products from most to least energy consumption per product. Table 13.2 looks at the nationwide energy consumption of these appliances. Like many of the home electronics products discussed in Chapter 12, many small household appliances use power supplies and/or battery chargers resulting in power losses regardless of whether they are in use, in charging mode, or supposedly "off."

TABLE 13.1 Electricity Consumption of Selected Miscellaneous End-Uses

Product	Average Annual Energy Use (KWh)
Pool Pump	3,000
Pool/Hot Tub Heater	2,300
Waterbed Heater	900
Water Dispenser (hot & cold)	827
Large Aquarium	550
Well Pump	400
Block Heater	250
Home Security System	195
Garage Door Opener	58
Rechargeable Tools	37
Coffee Maker	20
Answering Machine	19
Electric Shaver	11
Blender	7

Source: ECOS Consulting, 2006: *Final Field Research Report for the California Energy Commission;* U.S. Department of Energy, 2007, *www.eia.doe.gov*

TABLE 13.2 National Energy Use of Selected Miscellaneous End-Uses

Product	Number in Use (millions)	National Energy Use (billion kWh/year)
Pool Pump	6.5	9.8
Pool/Hot Tub Heater	3.3	7.6
Waterbed Heater	6.4	5.7
Well Pump	13.8	5.5
Home Security System	18.7	3.9
Large Aquarium	4.5	2.5
Ground Fault Circuit Interrupter (GFCI Outlet)	411.9	2.5
Rechargeable Tools	47.7	1.8
Garage Door Opener	28.4	1.7
Answering Machine	65.7	1.2
Coffee Maker	51.3	1.0
Water Dispenser (hot & cold)	1.1	0.8
Blender	84.3	0.6
Block Heater	2.3	0.6
Electric Shaver	50.3	0.6

Boost Energy Efficiency Whenever Possible — It's Easy On The Pocketbook And The Planet

As you consider energy-saving opportunities in the home, look over this list. If you have a lot of these or other miscellaneous energy users in the home, there may be opportunities for savings. Recommendations are listed below for reducing the energy use of a few of these household products.

■ Furnace Fans

Refer back to Chapter 4 on heating systems. If your furnace is improperly sized, or if the fan thermostat is improperly set, the fan may operate longer than it needs to. If you're getting a lot of cold air out of the warm-air registers after the furnace turns off, this may be the case. Along with making you uncomfortable, the fan is wasting energy. On the other hand, if the fan shuts off too soon, heat from the furnace will be wasted. Have a service technician check the fan thermostat setting if you're unsure.

■ Well Pumps

Well pumps are common in rural areas. The amount of energy they use is dependent on how deep the well is, the quality of the pump, and the pressure controls. If the pump seems to be switching on more often than it should, there may be a leak in the system or the pressure switch may not be functioning properly. Have the system checked out. You can also save on well-pumping electricity costs by reducing your water use (see discussion on water conservation in Chapter 6).

■ Swimming Pools

The energy used to operate the cleaning and filtering equipment of a typical pool for one swimming season can equal the amount of energy used to power the average home for the same period. In Sunbelt states, pool pumps can account for 25–30% of home electricity use! Many pool pumps are oversized; installing a replacement pump can save 40% of pumping energy use. Also, try reducing the amount of time you filter the pool each day. You may be surprised how much you can cut filter-

TABLE 13.3 Energy Savings from Pool Pump Upgrades

Upgrade	Energy Use (kWh/year)	Energy Cost ($/year)	Energy Savings
Typical Pool	3,000	285	—
Pump Replacement (Downsizing)	1,800	171	40%
Reduced Pump Run Time	1,200	114	60%
Combination of Above	720	68	75%

Assumes 9.5¢ per kWh.

Source: U.S. Department of Energy, 2007, www.eere.doe.gov

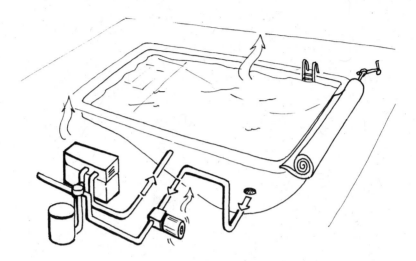

The pump, filter and heater needed to operate your pool can use a lot of energy. Improve energy efficiency by installing a high-efficiency heater and pool pump, using a pool cover, and managing water temperature.

ing time without negatively impacting the water quality. Table 13.3 above summarizes typical savings from these steps. Contact your utility to see if it offer rebates for efficient pool pumps.

Pool heaters are also a big energy user, costing as much as $1,000 per year or more to operate depending on the type of heater and how often you use it. Heat pump and solar pool heaters can cut your pool heating bills dramatically compared to electric resistance or gas models. Pool covers can also save energy by reducing heat losses. Other benefits include reduced evaporative water loss, chemical use, and cleanup time. Your state or utility may offer incentives for the installation of solar or heat pump pool heaters.

■ Spas and Hot Tubs
Spas and hot tubs can use a tremendous amount of energy. If you have one, keep it covered with a tight-fitting insulated cover when it's not in use. If installing a spa or hot tub, insulate it well around the sides and bottom.

For More Information:

For tips on reducing the energy used to heat and filter your swimming pool, check out the Department of Energy's online consumer resources. Look for "Swimming Pool Heating" under the Water Heating link.
www.eere.energy.gov/consumer/your_home/

■ Waterbeds

Approximately 5% of homes have waterbeds; most are heated with electric coils running underneath the mattress. Waterbeds can be the largest electricity user in the home — exceeding even the refrigerator and water heater! Simple habits can significantly reduce energy waste from waterbeds. Regularly making the bed with a comforter can save more than 30%; insulating the sides of the bed can save over 10%. You might also consider putting the waterbed heater on a timer so that it doesn't waste energy throughout the day.

■ Electric Blankets

Each electric blanket in a house uses an average of 120 kWh per year. If you use an electric blanket and frequently forget to turn it off in the morning, you can save energy by buying a simple timer control. Putting a second blanket or quilt over the electric blanket also saves energy, but be sure to check the electric blanket's instructions for possible precautions against this practice. Electric blankets may actually save energy by allowing you to turn your thermostat down further at night.

■ Block Heaters

If you live in a cold area and use a block heater to help your car start on cold mornings, you might be surprised at how much energy it draws. It is generally recommended that you plug in your vehicle for 2 to 3 hours prior to your departure time.

Rather than plugging in the car before you turn in for the night, use a timer or install an extra circuit with a timer-controlled switch in the house, and plug in the car when you get home. Then set the timer for 2 to 3 hours before you need to leave in the morning. If you have an irregular schedule, consider using a power-saver cord. A power-saver cord uses a built-in thermostat to sense the temperature of the coolant in the engine. If the temperature drops below the setpoint, power from the block heater is allowed to flow to your engine, ensuring that your vehicle will always be ready to go.

■ Small Appliances and Rechargeable Devices

While each of these devices uses relatively little energy on its own, most homes have several (if not one or two dozen!) plugged in at all times. This can add up to $25, $50, or even $100 or more each year. At the national level, this means more than $1 billion in annual energy costs and the associated pollution emissions. Many of these items waste energy through inefficient power supplies and battery charging systems. Others incorporate clocks or other features that require a constant power draw, even if the feature is not used — how many clocks do you need in one kitchen? Unplug these products when they are not in use or when their batteries are fully charged.

Use Your Common Sense

Saving energy with many of these incidental energy-consuming products — and with other products throughout the house — is very easy. Most of the time, it just takes some common sense. If we all become a little more aware of the energy we use, we might just begin to solve some of our major environmental problems, and we'll end up with a little more money in our pockets as well.

For More Information

In each chapter we have offered specific suggestions about where your next steps should be to find the products and services you are looking for. These resources are good to revisit, as policies, technologies, and product lines evolve. If you have additional questions or concerns that have not been answered in this book, or for the most up-to-date information, we encourage you to check out the resources listed below. This list includes key government agencies and research labs, leading newsletters, consumer and environmental organizations, trade associations, and private businesses offering products and services. More web resources can be located under "Energy Efficiency-Related Websites" on the ACEEE website (www.aceee.org/altsites/).

Government Resources

The U.S. Department of Energy (DOE) offers information to the public through the Energy Efficiency and Renewable Energy Information Center and Web site. The site includes information on getting a home energy audit, tips for saving energy throughout the home, guidance on using solar energy, suggestions for remodelers and apartment dwellers, and a buying guide for purchasing energy-efficient appliances. A toll-free number puts you in contact with professional staff that can assist you with questions about home energy use, renewable energy, recycling, and related issues.

Phone: (877) 337-3463
Website:
www.eere.energy.gov/buildings/info/homes/index.html

The U.S. Environmental Protection Agency (EPA) offers consumers a wealth of information on environmental quality at home and in your community. Information on local conditions, climate change mitigation, appliance recycling, chemicals and pollutants are all accessible from the home page.

Website: www.epa.gov
Hotlines: www.epa.gov/epahome/hotline.htm

ENERGY STAR. The U.S. EPA and DOE run the ENERGY STAR program. Consumers can call the ENERGY STAR hotline with questions or visit the Web site to search for ENERGY STAR-qualifying products by brand, type, size, efficiency level, and model number. The site also contains a number of tools and resources including a rebate finder and store locator, energy savings calculators, and how to find a contractor.

Phone: 888-STAR-YES
Website: www.energystar.gov

Lawrence Berkeley National Laboratory offers information on energy-efficient appliances, residential energy software, utility programs, home energy ratings, and financing options as well as numerous reports, case studies, and newsletters. The lab also offers an on-line interactive tool, the Home Energy Saver, to analyze your home for energy savings.

Website: www.lbl.gov/
Home Energy Saver: hes.lbl.gov

Your local energy office, utility or extension service may have additional information on energy savings, purchasing incentives, recycling programs and renewable power opportunities relevant to your particular location. You can usually access contact information for these agencies through your governor's office website.

Consumer and Environmental Organizations

The Alliance to Save Energy is a coalition of prominent business, government, environment, and consumer leaders. Energy-saving tips and information for consumers can be found at their website, as well as information for educators and the media.

Phone: (202) 857-0666
Website: www.ase.org/consumer

The Center for a New American Dream provides a wealth of re-sources aimed at helping American consumers make decisions that are more environmentally sustainable and socially just.

Phone: (877) 68-DREAM
Website: www.newdream.org

The Consumer Federation of America/Consumer Research Council provides information on ways to save energy and money throughout the home and answers to common questions about purchasing energy-efficient products and services.

Phone: (202) 387-6121
Website: www.buyenergyefficient.org

Environmental Defense maintains the Green Advisor website as a resource for consumers interested in making environmentally sound purchasing decisions. The site links to information and educational resources from a wide variety of environmental groups and offers tips on buying environmentally friendly products; cooking and eating a green diet; recycling and waste reduction; and green travel and recreation.

Website: www.greenadvisor.org

The Natural Resources Defense Council (NRDC) is a nonprofit organization dedicated to protecting the world's resources and ensuring a safe and healthy environment for all. Their website provides information about the connection between energy use and the environment, clean power sources, air pollution, and more.

Website: www.nrdc.org

Trade Associations

The Association of Home Appliance Manufacturers (AHAM) provides information on a wide range of appliances through its "Just for Consumers" website. The site also offers advice from the experts on appliance purchasing, use, maintenance, and repair.

Website: www.aham.org/consumer/

The Gas Appliance Manufacturers Association (GAMA) is an association of home heating and water heating appliance and equipment manufacturers. GAMA hosts a gateway to a database of certified products where consumers and contractors can look up certified equipment according to size, fuel, efficiency specifications.

Website: www.gamanet.org

The Consortium for Energy Efficiency (CEE) is a nonprofit public benefits corporation that brings together North American utilities, environment groups, research organizations and state energy offices to develop initiatives that encourage energy-efficient products in the marketplace. These initiatives often include "better than ENERGY STAR" ratings for a number of home appliances and products.

Website: www.cee1.org

Directories of Products and Services

EnergyGuide sells a growing list of high-quality products to help reduce your energy costs along with the information you need to guide your decisions. Enter your zipcode and the site will determine what, if any, utility rebates apply to your locale and apply the rebate directly to your purchase. The site can also help consumers find a contractor or select an energy supplier.

Website: www.energyguide.com

Energy Federation Incorporated (EFI) offers a wide variety of water and energy saving products in their on-line catalog. The site will apply any applicable utility rebates directly to your purchase.

Website: www.efi.org

GAIAM Real Goods has been a leader in the sustainable products market offering goods, services, and advice on saving energy and using renewable energy in the home for more than 20 years. Over 10,000 products ranging from air and water purification devices to clothing, lighting, batteries, and toys are available on its website.

Phone: (800) 762-7325
Website: www.realgoods.com

Oikos Green Building Source, hosted by Iris Communications, is a website offering information on energy-efficient products and a database of 1,900 companies that make or distribute them. The site also has an on-line catalog of books, videos, and software for sustainable, environmentally sound construction.

Phone (for free print catalog): (800) 346-0104
Website: www.oikos.com

Home Energy and Environmental Publications and Newsletters

Consumer Reports, published by the Consumers Union, is a valuable resource for information on a range of products, including home appliances and consumer electronics. Product reviews provide ratings of key features, performance, repair history, and pricing, in addition to information on energy use and efficiency. A web-based version of the magazine, along with an archive of product evaluations and reports, can be accessed for a modest fee. A new online publication called *Greener Choices* has a number of additional ratings and tools aimed at the environmentally sensitive consumer.

Website: www.consumerreports.org
www.greenerchoices.org

BuildingGreen is an independent company that provides information to building-industry professionals and policy makers. Their reports improve environmental performance and reduce the adverse impacts of buildings. They also publish *Environmental Building News*.

Website: www.buildinggreen.org

The Green Building Resource Guide is a database of green building materials and products for design and building professionals. The database includes a Price Index Number, a factor which compares the cost of a green product to a product labeled Price Index Standard. The index assists architects in making decisions with respect to product specifications and project budgets.

Website: greenguide.com/

GreenerCars.org informs consumers of the greenest vehicles on the market. They provide a green score for every car and truck on the market, facilitating comparisons for car-buyers.

Website: www.greenercars.org

This Green Life, a free online journal of the Natural Resources Defense Council, includes articles on a range of domestic energy use and environmental issues and debated topics.

Website: www.nrdc.org/thisgreenlife/

Grist Magazine reports global environmental news with a sense of humor. *Grist* is a nonprofit organization, but their website and email services are free to the public.

Website: www.grist.org/

Index

ACCA (see Air Conditioning
  Contractors of America)
AFUE (see Annual fuel utilization
  efficiency)
Air conditioners (see also Cooling
  systems), 99–123
  basics of, 107
  central, 107–108
  dehumidification, 120
  ductless mini-splits, 111
  hot-dry climate, 123
  room, 109
  state of the art, 123
Air Conditioning Contractors of
  America (ACCA), 115
  ACCA Manual J, 84, 114, 117
Air distribution (see also Forced air
  system, Ducts), 47–48
Air exchange (see also Ventilation),
  46, 56–57
Air filters, 47, 48, 54–55
  check light, 118
  cleaning or replacement, 54,
    96, 122
  electrostatic (electronic,
    electret), 54–55
  location of, 55
  mechanical, 54
Air handler (furnace fan), 47, 63,
  86, 108, 218

efficiency, 86
energy use, 218
fan thermostat, 89, 218
fan-only switch, 118
variable speed fan, 86
Air leakage, 13, 16, 17, 19–23,
  30–35, 42, 46–48, 56, 73, 89
  caulking, 19–24, 32–35
  common sites, 19–23
  in ducts, 47–48, 73, 89
  importance of reducing, 15–19
  weatherstripping, 32–33
  in windows, 30–35, 42
Air quality, 16, 47–59, 81, 163
American Society of Heating,
  Refrigeration, and Air
  Conditioning Engineers
  (ASHRAE), 100
Andersen Windows, 39
Annual fuel utilization efficiency
  (AFUE), 11, 65, 71, 73–76, 79,
  84–87, 135
  defined, 65
  recommended levels, 84–85
  savings from improvements in,
    73–74
Aquarium, 215–217
Aquastat, 66, 89–90, 97
Argon gas–filled windows, 4, 38, 41
Asbestos, 77, 94

About the Authors

Jennifer Thorne Amann is on staff at the American Council for an Energy-Efficient Economy. She has written numerous articles and reports on energy use in buildings, appliances, and consumer products. She also has experience educating citizens on a range of environmental and consumer issues.

Alex Wilson is the founder and president of BuildingGreen, Inc., a 20-person information company that has served the design and construction industry since 1985. He is executive editor of *Environmental Building News*, a highly regarded newsletter on environmentally sustainable design and construction, and is co-editor of the *GreenSpec® Product Directory*. He is also the author of *Your Green Home* (New Society Publishers, 2006), and has written extensively on energy and building technology for such magazines as *Fine Homebuilding, Architectural Record, Journal of Light Construction, Landscape Architecture,* and *Popular Science*. He served for five years on the board of the U.S. Green Building Council and is currently a trustee of The Nature Conservancy - Vermont Chapter.

Katie Ackerly conducts research on building efficiency for the American Council for an Energy-Efficient Economy. She has produced a number of reports on residential and commercial building technologies. She also helps manage ACEEE's online consumer resources and has several years of experience as an artist and illustrator.

About ACEEE

The American Council for an Energy-Efficient Economy (ACEEE) is an independent, nonprofit organization dedicated to advancing energy efficiency as a means of promoting both economic prosperity and environmental protection. Support for our work comes from a wide range of foundations, government organizations, research institutes, utilities, and corporations.

ACEEE produces guides for consumers and procurement officials buying equipment and vehicles, including the *Consumer Guide to Home Energy Savings* and ACEEE's Green Book® Online (greenercars.org), an environmental guide to cars and trucks (updated annually).

The *Consumer Guide to Home Energy Savings* is accompanied by a condensed online version that contains more frequently updated information on how to find the most efficient home appliances and equipment.

Website: www.aceee.org/consumerguide

To find out more about ACEEE Publications, visit www.aceee.org/pubsmeetings, or order a publications catalog by contacting:

ACEEE
1001 Connecticut Avenue, N.W., Suite 801
Washington, D.C. 20036
Phone: (202) 429-0063 Fax: (202) 429-0193
E-mail: aceee_publications@aceee.org

If you have enjoyed *Consumer Guide to Home Energy Savings*
you might also enjoy other

BOOKS TO BUILD A NEW SOCIETY

Our books provide positive solutions for people
who want to make a difference. We specialize in:

**Environment and Justice • Conscientious Commerce
Sustainable Living • Natural Building & Appropriate Technology
Ecological Design and Planning • Educational and Parenting Resources
Nonviolence • Progressive Leadership • Resistance and Community**

New Society Publishers

ENVIRONMENTAL BENEFITS STATEMENT

New Society Publishers has chosen to produce this book on Enviro 100, recycled paper made with **100% post consumer waste**, processed chlorine free,
and old growth free.

For every 5,000 books printed, New Society saves the following resources:[1]

24	Trees
2,189	Pounds of Solid Waste
2,408	Gallons of Water
3,141	Kilowatt Hours of Electricity
3,979	Pounds of Greenhouse Gases
17	Pounds of HAPs, VOCs, and AOX Combined
6	Cubic Yards of Landfill Space

---

[1]Environmental benefits are calculated based on research done by the Environmental Defense
Fund and other members of the Paper Task Force who study the environmental impacts of the
paper industry.

For a full list of NSP's titles, please call **1-800-567-6772** *or check out our website at:*

www.newsociety.com

NEW SOCIETY PUBLISHERS